Military English

Edited by Kyutaek Cho

JUNG LIM PUBLISHER

PREFACE

머리말

이 책은 군사영어를 체계적으로 익힐 수 있도록 편집했다. 최근 대학에 군사학 관련학과나 교과목이 급속하게 팽창하고 있지만, 학생들에게 쉽고 유익하게 다가갈 군사영어 교재가 없음을 안타깝게 생각해 왔다. 이에 엮은이는 군사영어와 일반영어를 모두 아우를 수 있는 종합교재를 검토하게 되었다. 이 책은 오랜 기간 군사영어를 가르치면서 학생들에게 꼭 필요하다고 생각했던 엮은이의 의지의 실현이다.

20년 전 엮은이가 해군영어교관으로서 가르쳤던 경험이 군사영어 교재를 만드는데 바탕이 되었음은 물론이다. 엮은이는 이 책을 공부하는 모든 학생들이 군의 간부로서 영어와 관련된 문화 · 교양의 차원까지도 터득할 수 있게 만들었다. 군의 간부는 계급에 걸 맞는 신사로서의 행동과 지식을 겸비해야 한다. 그러므로 군사학과나 부사관과 학생들이 장차 장교나 부사관으로서 익혀야 할 내용으로 자료 수집을 했으며, 철저한 검정을 통해 군사영어라는 책으로 태동하게 되었다. 이 책은 학생들이 쉽게 공부할 수 있도록 다음과 같은 기준을 설정하고 있다.

첫째, 국군방송에서 방영했던 군사영어 가운데 영화로 제작된 부분을 활용해서 학생들이 입체적으로 학습하게 했다. 이는 실무에서 영어에 자신감을 키워주게 된다.

둘째, 본문의 내용을 장문으로 구성하여 3사관학교 진학시험에 대비한 독해력 향상을 가능하게 했다. 이는 장차 유학을 고려하는 간부들에게도 반드시 필요한 학습법이 된다.

셋째, 본문을 군과 관련하여 교양·문화를 아우를 수 있는 내용으로 편집했다. 또한 본문관련 사진이나 그림을 배치하여 학습자의 이해를 증가하도록 했다.

넷째, 어휘력 향상을 위해 매 과마다 어휘(Vocabulary Checkup) 코너를 마련하여 활용하게 했다.

다섯째, 숙어를 활용한 문장이해를 넓히기 위해 주요표현(Key Expressions)을 매 과마다 배치하고 있다.

여섯째, 매 과의 끝 부분에 재치와 교훈을 통해 교양인으로 거듭날 수 있도록 유머영어나 영시와 우리시를 배치하여 여유를 갖게 했다.

일곱째, 부록에 군사약어와 군사용어를 우리말과 영어로 배치하여 실용적인 군사영어를 활용할 수 있게 배려했다.

이상과 같은 책의 특성을 잘 활용하여 이 책으로 공부하는 모든 분들이 **군사영어**에 친숙해져서 소기의 목표를 달성할 수 있기를 기대한다.

끝으로 국군방송의 군사영어 제작진들에게 감사와 경의를 표한다.

2010년 1월

엮은이 씀

CONTENTS

목 차

Lesson 1.

Organization and Positions in the Military

Watching Movie

WE WERE SOLDIERS I

Today's Clip

Welcome to the new cavalry. We will ride into battle. And this... will be our horse.

You don't have to catch it. You don't have to feed it. But **I can assure you... that** neither the new technology nor your status as officers... will keep you above the danger.

Vocabulary Checkup

cavalry	공수부대
battle	전투
feed	먹이를 주다

technology 기술

status 지위, 계급

Key Expressions

- Welcome to the new cavalry. We will ride into battle.
- But **I can assure you... that** neither the new technology nor your status as officers... will keep you above the danger.

Let's Review!

A: Did you see the movie called "We WERE soldiers"?

B: Yes, a few months ago.

A: How was it?

B: I won't tell you the story, but **I can assure you that** it's a must-see.

WE WERE SOLDIERS II

Today's Clip

But this I swear... before you and before Almighty God... that when we go into battle... I will be the first to set foot on the field... and I will be the last to step off. And **I will leave no one behind.** Dead or alive... we will all come home together. So help me God.

Vocabulary Checkup

swear 맹세하다

almighty 전지전능한

set foot 발을 들여놓다.

step off 발을 빼다.

alive 살아있는

Key Expressions

- I will be the first to set foot on the field... and I will be the last to step off.
- And **I will leave no one behind.** Dead or alive... we will all come home together.

Let's Review!

A: Sir! We have to board the chopper!

B: We can't leave without Sergeant Smith!

A: Sir! We can come back for him later!

B: No! **I will not leave anyone behind!** We must get him and bring him with us!

READING

Good afternoon everyone,

First of all, thank you for inviting me today to talk to you about careers in the Military. There are lots of opportunities for you to get a good career in many different parts of the Military, like the Army, Navy and Air force, for example. Today, I'd like to tell you about a career as a Non-commissioned Officer or NCO, as we call it.

To become an NCO, you must first have very good results in general education and pass all the necessary school leaving exams. After that you can join the Military and you must successfully complete 12 weeks basic training. During basic training, you'll learn about weapon handling, field-craft, nuclear, chemical and biological warfare and many other military subjects. You must also be very fit, because during basic training you must run at least 20 kilometres a day for the first three weeks without carrying any equipment. After that you will go on shorter runs, but you will carry your military equipment, and your military equipment can weigh up to 30 kilos.

A Tank

A Jeep

At the end of basic training, you will have lots of different tests and exams, and if your grades are good enough, you will go to NCO and technical training school to learn how to be a successful NCO. This training takes another 3 months and if you successfully complete that, we will send you to a regiment as a lance-corporal to practise what you have learned. Eventually, you could become a Regimental Sergeant Major or become an officer.

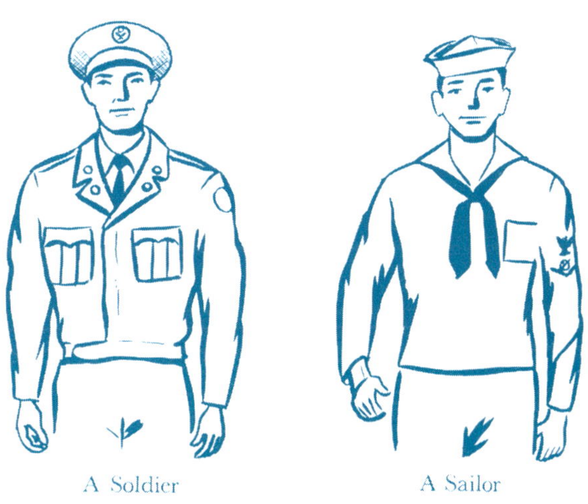

A Soldier　　A Sailor

There are many, many opportunities to travel in today's military. As many of you know, there are many areas all over the world which need the protection of peace keepers, and these peace keepers are provided by many different countries from the United Nations. Years ago, soldiers were trained only for combat, but nowadays, the job they do is much different. They are trained to take care of people who can't take care of themselves. After a civil war, such as in the former Yugoslavia, many thousands of people are left without homes, food and basics. It is a peacekeeper's job to help them find homes, food and other things that can help them have a normal, peaceful life.

So everyone, I hope you can see that a life in the military can help world peace and make sure that innocent people are not hurt by others that simply want to take their land, and stop them living peaceful lives. How are any Armed Forces of the U. S. Military comparing with the U. S. Navy?

The United States Marines

The United States Marines are part of the Navy. They are fighting our enemies in many far-away lands. The Marines fight on the sea, on land, and in the air. Navy landing craft take them to the sea coasts of enemy lands. Many times the Marines are the first men to land of the beaches. They take guns, and shells, and tanks ashore with them. They fight to win the land from the enemy and hold it.

Like sailors and Coast Guard men, the Marines go to school before they fight. They learn gunnery and how to handle all kinds of firearms. They learn how to handle tanks. They learn how to fight in any land at any time, day or night, with guns, with knives, or with their hands. The Marines can fight any enemy to the finish. They are real fighting men. You can tell a United States Marine by the badge he wears and by the way he dresses. He is not dressed like a sailor or Coast Guard man.

The Coast Guard

The Coast Guard is part of the United States Navy when there are enemies to fight. Men of the coast Guard man ships and other craft. Some of the ships are called Coast Guard cutters. The cutters help to keep our ships safe from enemy subs and planes. There are many Coast Guard men and officers on landing craft, too. They help to make landings on enemy coasts. The coast Guard is ready for any job at sea.

Like other sailors, Coast Guard men go to school to learn about ships and guns. They learn how to march and how to handle firearms. They learn how to save lives, too.

Coast guard men guard our beaches at home. They fight on the sea, and on the coasts of lands that are far away.

You can tell a Coast Guard man by the badge he wears on his arm. The men are proud to wear this badge. They are proud to be Coast Guard men.

The Army

The Army is the largest of the services. Army men are at bases all over the world. Army men are called soldiers. Some soldiers march and carry rifles. Some ride in tanks. Some are para-troopers. They all are trained to fight.

The Army is one part of the service team. Army men are proud of their outfit just the way that sailors are proud of theirs.

The Air Force

The United states Air Force is the third big part of the Armed Forces team. Remember that marines are part of the Navy. The Coast Guard during wars becomes part of the Navy. At other times the Coast Guard is part of the Treasury Department. The U. S. Air force is not the only force also has them. And the Army and the Marines have many planes and helicopters which they use for their own jobs. The U. S. Air Force has the largest bombers and most of the land-bases in this country and other lands.

Air Force men are called airmen. Air Force officers have ranks like those of Army and Marine Corps officers. They are generals instead of admirals, and colonels instead of captains as in the Navy. Thy Air Force, like the Army and Marine Corps, also has sergeants. However, instead of privates and corporals, the Air Force has Airman Third class, Airman Second Class, and Airman First Class. An Airman Third Class is equal to a Seaman Recruit in the Navy and a Private Recruit in the Army. An Airman Second Class is equal to a Seaman Apprentice in the Navy and a Private First Class in the Army. An Airman First Class is equal to a Seaman in the Navy and to a Corporal in the Army.

I. Military Ranks

A. Noncommissioned Officer

원 사	Sergeant Major(SGM)	
상 사	Master Sergeant(MSG)	
중 사	Sergeant First Class(SFC)	
하 사	Staff Sergeant(SSG)	

B. Company grade officer

대 위	Captain(CPT)	
중 위	First Lieutenant(1LT)	
소 위	Second Lieutenant(2LT)	
준 위	Warrant Officer(WO)	

C. Field grade officer

대 령	Colonel(COL)	
중 령	Lieutenant Colonel(LTC)	
소 령	Major(MAJ)	

D. General grade officer

대 장	General(GEN)	
중 장	Lieutenant General(LG)	
소 장	Major General(MG)	
준 장	Brigade General(BG)	

II. Unit Size

육 군		해 군		공 군	
군	Army	함대	Fleet	방공포 사령부	Air Defence Artillery Command
군단	Corps	전단	Flotilla	전투비행단	Fighter Wing
사단	Division	전대	Squadron	전대	Squadron
여단		Brigade			
연대		Regiment			
대대		Battalion			
중대		Company			
소대		Platoon			
반		Section			
분대		Squad			

III. Positions

참모총장	Chief of Staff
사 령 관	Army Commanding General
군 단 장	Corps Commanding General
사 단 장	Division Commanding General
여 단 장	Brigade Commander
연 대 장	Regiment Commander
대 대 장	Battalion Commander
중 대 장	Company Commander
주임원사	Command Sergeant Major
행정보급관	First Sergeant
소 대 장	Platoon Leader
부소대장	Platoon Sergeant
분 대 장	Squad Leader
반 장	Section Leader
팀 장	Team Leader

IV. Military Occupational Specialty

보 병	Infantry
포 병	Artillery
기 갑	Armor
공 병	Engineer
통 신	Communication
화 학	Chemical
보 급	Supply
병 기	Ordnance
수 송	Transportation
헌 병	Military Police
항 공	Aviation
정 훈	Troop Information and Education
법 무	Judge Advocate
군 종	Chaplain
의 무	Medical
기 무	Security
경 리	Budget

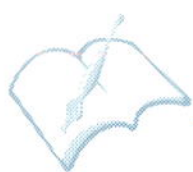

Humor English

Too much fee

"Are you a lawyer?"

"Yes"

"How much do you charge?"

"A thousand dollars for four questions."

"Isn't that awfully expensive?"

"Yes, What's your fourth question?"

너무 과다한 수수료

"당신이 변호사이신가요?"

"그렇습니다."

"수수료가 얼마인가요?"

"질문 네 개에 1천 달러입니다."

"그건 너무 비싼 거 아닌가요?"

"그렇습니다. 당신의 네 번째 질문은 무엇인가요?"

Lesson 2.

Navy Life

Watching Movie

U-571 I

Today's Clip

- Mr. Tyler, call the crew to quarters.
- Aye, Captain. Quarters! Everybody fall in! Let's get in formation now! Let's go! Now! Formation! Captain Dahigren, all crew members present and accounted for, sir!
- Very well. At ease. I'm sorry your liberty's been cut short, but **our schedule's changed**. We have two hours to get stowed for sea and set the maneuvering watch.

Vocabulary Checkup

quarters 막사, 병영

fall in 집합!

get in formation	정렬하다
cut short	단축하다, 갑자기 끝내다
at ease	쉬어!
stow	(화물칸 등에) 짐을 싣다.
maneuvering watch	작전 시계

Key Expressions

- I'm sorry your liberty's been cut short, but **our schedule's changed.**
- We have two hours to get stowed for sea and set the maneuvering watch.

Let's Review!

A: OK everyone. **Our schedule is changed.**

B: What? Why? To what?

A: We are not meeting at the classroom. We will meet at the conference room, as the building the classroom is in will be locked that day.

B: Makes sense.

U-571 II

Today's Clip

- More soup, Captain?
- Thank you.
- I wish I was going with you, gentlemen.
- **I feel the same way** you do, Carson.
- Gentlemen, **I'd like to propose a toast**. To the officers and crew of the S-33... and our distinguished guests, Mr. Hirsch and Mr. Coonan. Success, gentlemen!
- Success!

Vocabulary Checkup

propose	제안하다
toast	건배
distinguished	뛰어난

Key Expressions

- **I feel the same way** you do, Carson.
- Gentlemen, **I'd like to propose a toast**.

A: **I'd like to propose a toast.**

B: To what?

A: Jenny got promoted to first lieutenant.

B: Good news!

A: So, to the new first lieutenant, Jenny! Cheers!

B: Cheers!

Recruit Days in the Navy

Marlinspike Seamanship

The recruits are learning to tie knots. They learn to make hitches.

All these things are part of what is called marlinspike seamanship.

Sailors must know all these things. They are very important.

Navy men work around boats and lines all the time. They have to be good seamen.

One sign of a good seaman is his skill with knots and hitches.

The Wheel and the Compass

The Chief is showing Sam and his buddies how to use the wheel and the compass.

They learn about the engine order telegraph, too. All these things are on the bridge of a ship.

The man who steers is called the steersman.

When he turns the wheel to the right, the ship goes to the right, or starboard.

When he turns the wheel to the left, the ship goes to the left, or port. The compass shows him the course he is steering. The steersman never changes course without being told.

Sam and his buddies are learning fast. They will be good sailors.

Anti-Aircraft Guns

Today the recruits are being shown how to work an anti-aircraft gun. The chief shows them how to load it. He shows them how to fire it. He shows them how to move it up and down and from side to side.

The recruits learn all about the duties of a gun crew.

They learn that the pointer moves the gun up and down. They learn that the trainer moves the gun from side to side.

The recruits learn all about the duties of a gun crew.

They learn that the pointer moves the gun up and down.

They learn that the trainer moves the gun from side to side.

They learn what the first loader does. All these are important jobs. So are all the others in a gun crew. Everyone is important when the guns are needed.

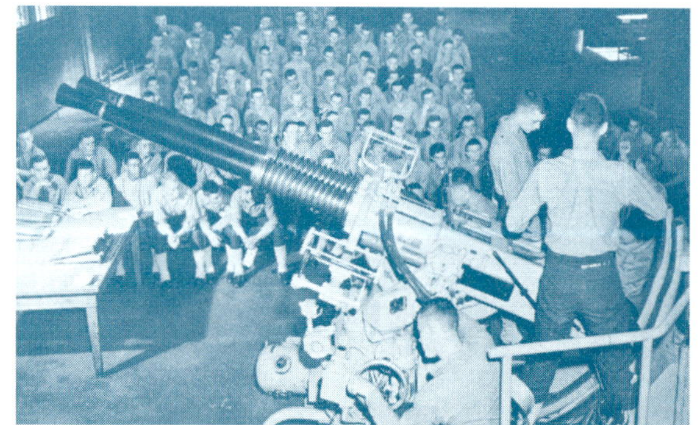

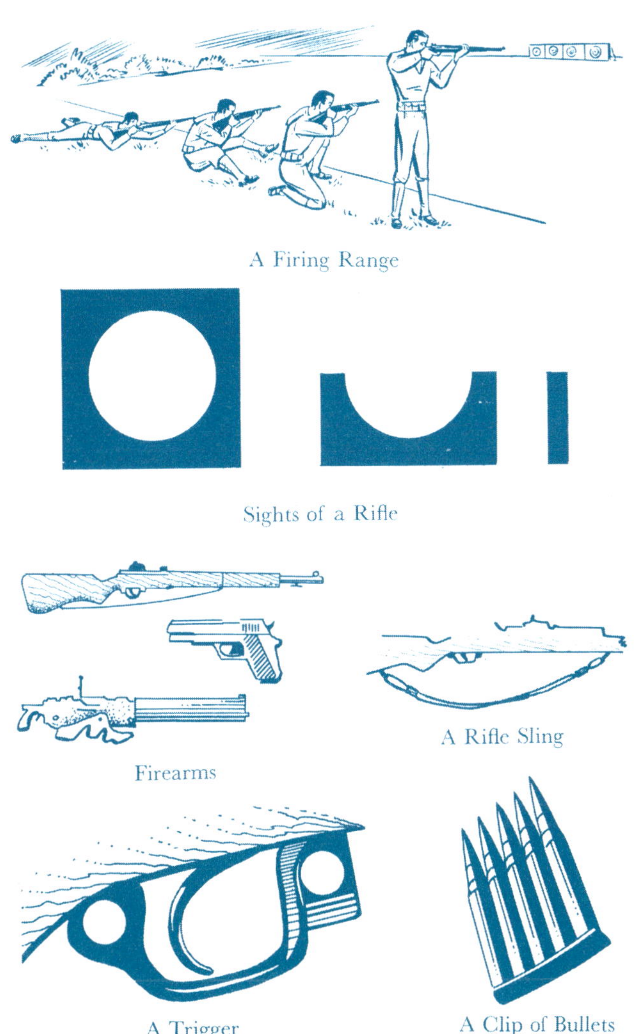

A Firing Range

Sights of a Rifle

Firearms

A Rifle Sling

A Trigger

A Clip of Bullets

Using the Rifle

Sam is learning to use a rifle. He is getting ready to go on the firing range.

The firing range is a place where there are many targets to shoot at. Sam will shoot at a target. Another man will mark the spot where Sam's bullet hits. Then Sam will know how well he shoots.

There is a lot to learn about a rifle. A man must learn how to use the strap under the rifle. This strap is called the sling. A man has to learn to adjust the sights.

The sights help him aim a rifle. A rifle is a firearm. There are other firearms, too, such as pistols, shotguns, and small machine guns.

Rifle Drill

Sam has just been given the order "Left shoulder, arms." This is part of rifle drill. Rifle drill is called the Manual of Arms.

The Chief is showing Sam what to do. There is a right way and a wrong way to handle a rifle. The Chief wants Sam to be right.

Rifle Exercise

Everybody needs exercise to stay healthy. One of the exercise recruits get is with a rifle. They hold their rifles out straight in front of them. They raise them over their heads. They step to one side. The rifles weigh about nine pounds each. When the men toss these around they get good exercise. Also, they get used to the feel of the rifles. This makes them better able to handle a rifle. It does not feel strange to them. Sam likes exercising with a rifle. He has learned the right way to do it. He is fast and careful. He is getting to be a good sailor.

THINGS TO TALK ABOUT

1. What do we mean by "marlinspike seamanship?"
2. Why must sailors be good seamen?
3. Tell about the duties of the steersman of a ship.
4. Tell about some of the men who stand watch on the bridge of a ship.
5. Tell about the duties of an anti-aircraft gun crew.
6. What do we mean when we say that everyone is important when the guns are needed?
7. Tell what you know about firing a rifle?
8. Tell what you know about the manual of Arms.
9. Why is rifle exercise good exercise?
10. Tell what you know about the firing range.

Learning About Ships

The Chief is showing the recruits the different parts of a ship. He is using a ship model. He shows them the bridge and different sections of the deck. He shows them models of many different kinds of ships.

There is a lot to learn about a ship. Sam tries hard to learn all he can.

The Chief shows them a model destroyer. This is a small, fast ship. He shows them a model of a tanker. This is a big, slow ship that carries oil. The chief shows them a model of a submarine.

This is a ship that can go under the water. He shows them a model of a carrier. This is a ship that carries aircraft.

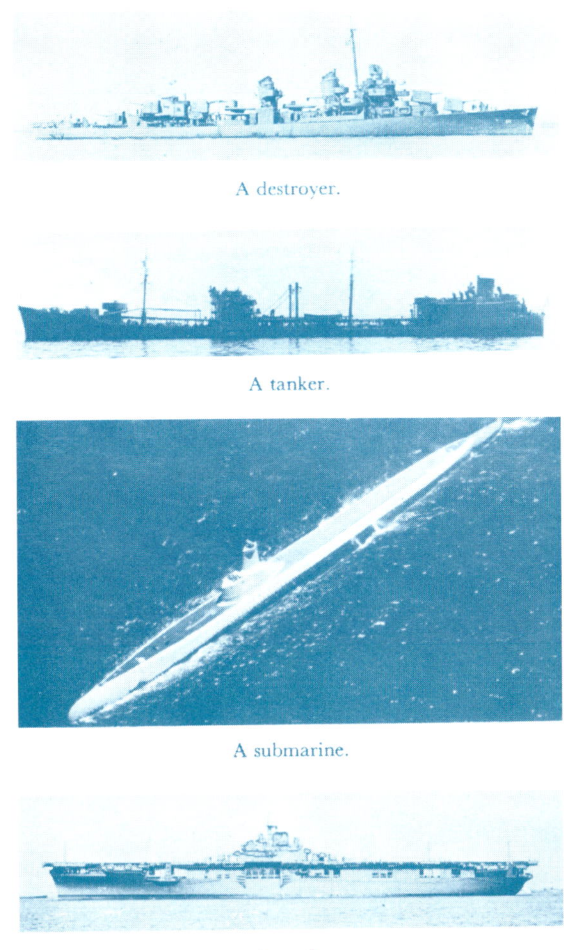

A destroyer.

A tanker.

A submarine.

A carrier.

Company Honor Man

Sam has been chosen Company Honor Man. The company votes to choose its honor man. To be picked, a man must have worked and studied hard. He must be tops in company spirit. This means he always works for the good of the company.

Sam did all these things. He tried to do his best all through training. He tried to make the company a good one. All the recruits knew this, which was why they voted for Sam. They knew he had earned it. It is a real honor to be chosen company honor man.

Graduation Review

This is the day the recruits have been working for! Today is graduation day. All the companies will pass in review. This means they will march past the reviewing stand.

In the reviewing stand will be the Captain and his officers. There may be an Admiral there, too. All around the reviewing stand will be civilian guests. There will be the families of many of the recruits. Their girl friends may be there, too.

Everyone is proud of the new sailors. They have earned their way through boot camp. Now they are ready for the Fleet. They will go to ships, or schools, or bases. A whole new life lies ahead of them. They are proud sailors.

Last Salute

Sam is holding his rifle at "Present Arms." This is his last day as a recruit. Holding his rifle this way is a kind of last salute.

Sam is thinking back over the days at boot camp. He is almost sorry to leave. He worked hard, and he learned a lot. He has gained weight and is a lot stronger than he used to be. Recruit training has been good for him.

Sam also thinks of the days ahead. He wants to go to a ship. He has learned a lot. Now he is ready to be a sailor. He will be a good sailor.

THINGS TO TALK ABOUT

1. Tell about some different kinds of Navy ships.
2. What are the "turrets" in a ship?
3. Tell what you know about the Marines.
4. What do you know about Marine boot camp?
5. Tell what you know about the Coast Guard.
6. What is a Coast Guard cutter?
7. Tell what you know about the Army.
8. What is graduation review?
9. What do we mean when we say that the hat the have "earned their way through boot camp?"
10. Tell about some of the duties a sailor can go to when he leaves boot camp.

Lesson 3.

LT.(jg) John F. Kennedy

Watching Movie

Pearl Harbor I

Today's Clip

- The mission I'm asking you to **volunteer for** is exceptionally dangerous.
- **It's a good bet** that in the next six weeks... you or he... will be dead.
- Everyone brave enough to accept this... step forward.

Vocabulary Checkup

mission	임무
volunteer	지원하다
exceptionally	매우, 예외적으로
a good bet	가능성이 높은 일

Key Expressions

- The mission I'm asking you to **volunteer for** is exceptionally dangerous.
- **It's a good bet** that in the next six weeks... you or he... will be dead.

Let's Review!

A: Sergeant Jackson. **Take a look at** my rank and see if it was sewn on right.

B: Hey, that's right! You just got promoted!

A: Just last week. I'm not sore of this is on correctly though.

B: Hmm... Come here.

A: Let me **take a look at** that...

Pearl Harbor II

Today's Clip

- Japs transmitting! Japs transmitting!
  Captain, radar reports Japanese patrol boats. Range 400 yards.
- How many yards?
- 400 yards and closing, sir.
- Richards, Japs! 400 yards! Let's go!
- Admiral Halsey's **ordering the cruisers to open fire!**
- We gotta sink these Jap boats before they give us away the position. Go to general quarters!
- Go! Go! Go! Go! Go! Go! Go! Go!
- Oh, my God.
- This can't be good!

Vocabulary Checkup

transmitting	송신
patrol boats	초계정
to open fire	발포하다
general quarters	전원배치

Key Expressions

- Admiral Halsey's **ordering the cruisers to open fire!**
- We gotta sink these Jap boats before they give us away the position. Go to general quarters!

Let's Review!

A: General Kim was looking for you. Did you do the paperwork that he asked you to do?

B: Oops. I forgot.

A: Oh, no. **This can't be good.**

READING

B. LT.(jg) John F. Kennedy

In 1943, during the Second World War, John Kennedy was a Lieutenant (jg) in the United States Navy. He was the communications officer of a PT boat. A PT boat is a small boat and it can move very fast. It is about 77 feet long and is heavily armed.

LT.(jg) JOHN KENNEDY

Lt Kennedy's boat, PT-109, patrolled[1)] waters between islands on which enemy soldiers were still active. The enemy was trying to move their troops farther north. The mission of the PT patrols was to stop the movements of enemy troops and to tell headquarters of enemy actions.

1) **to patrol**: To watch for the enemy.
They **patrolled** the river every day but they did not find the enemy.

One night, while PT-109 was on patrol, an enemy ship appeared. It approached PT-109 at great speed. Before PT-109's 37-millimeter gun could be fired, the boat was hit by the enemy ship. PT-109 was cut in two by the enemy ship. The back half of the boat went with the enemy ship. The front half of the boat with all 11 crew[2)] members[3)] on it remained floating. The front half of the boat was surrounded by its own oil.

When daylight came, the men saw an island not fat away. But the island had an enemy base on it. Lt. Kennedy gave orders to try to reach another small island, about 3 miles away.

When PT-109 did not return after its night patrol, the squadron headquarters thought the boat and its crew were lost. The men at the base did not know that 11 men from PT-109 were still alive and that many of them were injured.

The crew left the remaining half of PT-109. They took the big piece of wood holding the 37-mm gun and used it as a boat. The crew hung on to it, except one man, an engineer who had been severely burned. Lt Kennedy himself pulled this man through the water. That evening, more than 4 hours later, they reached the small island. They were very tired and very hungry.

2) **crew**: A group of men working together under one officer.
The Captain told the ship's crew that a storm was coming; the **crew** hurried to their stations.

3) **member**: A part of a whole; a person belonging to a group.
He is a **member** of the U. S. Air Force; he is a **member** of an airplane crew.

But Lt Kennedy did not rest. To the surprise of his men, he went back into the sea again. He began to swim. The men knew that Lt Kennedy was not in good condition. They knew that the ocean currents[4] were very strong. They also knew that some of the fish in these waters could bite off a man's legs.

After swimming for an hour, Lt Kennedy reached a point from where he hoped to use his light to signal other to siPT boats. He knew that boats from his squadron often moved on night patrol near this point. He waited in the water for about 2 hours. No PT boat came.

Into the boats.

Learning to handle the oars.

He started to return to the island. But a strong current held him back and carried him away. Twelve hours later he finally reached the beach[5] where his friends were. He was very tired and he was sick with a high fever.

4) **current**: Movement of the water.
The **current** of the water carried the body away from the boat.

5) **beach**: Shore along lakes, seas, or oceans, usually covered with sand or pebbles.
I like to relax on a sandy **beach** in the hot sun.

Since no food or water could be found on the small island, Lt Kennedy decided that there was only one thing to do. The crew would have to swim to an island that was closer to the usual PT patrol passage[6]. Once again they set out, using the big piece of wood as a boat. Once again, Kennedy pulled the engineer behind him.

After 3 hours, they reached the island. They found no food, only coconut[7] milk to drink. That night it rained heavily.

The men drank the water from the leaves of trees. The water tasted bad. The next morning the men realized why the water tasted bad. There were many birds in the trees. The birds made the trees very dirty. This event led to their naming the place Bird Island.

They continued to search for food and water. They still hoped to signal a PT boat as it came through the passage on night patrol. Finally, they decided someone would have to swim to the larger island to obtain food; although they knew they would probably find enemy troops. Kennedy asked Ensign Ross to go with him on this mission. The two men swam for over an hour. When they landed on the beach, the coral[8] cut their bodies. Instead of soldiers, they found a broken enemy landing boat on the beach. They examined it and came across a container of water and pieces of old bread. They ate the bread hungrily.

6) **passage**: The way of going from one place to another; the usual way into or out of a particular place.
We went through the **passage** to the next street.
The ship's **passage** through the rocks was a dangerous one.
7) **coconut**: A fruit with a very hard shell.
8) **coral**: Hard, sharp material growing in the water.

That night the two men waited for U. S. patrol boat to come through the passage. None came. When the sun rose, they found a one-man canoe[9)] under some trees. There seemed to be no boats approaching the passage during the second night. Lt Kennedy took the canoe and went to the smaller island. He took with him the few provisions they had found. The old bread was the first food they had eaten since the enemy ship hit them.

On the way back to Nauru Island, Kennedy met more trouble. Turbulent weather[10)] rose and a strong storm followed. His canoe was turned over by the strength of the water. He was thrown about in the sea. A native[11)] found him and pulled him by canoe directly to where Ross was waiting-surprising Ross who thought he was well hidden.

Lt Kennedy was in very poor condition; it was difficult for him to breathe. When he was able to talk at last, the two men tried to speak with the natives, but the natives did not understand English. Lt Kennedy said "Rendova, American" several times.

9) **canoe**: A long, light narrow boat.
10) **turbulent weather**: Bad weather. Rain or wind.
The turbulent weather kept the boats from going to sea.
11) **native**: Local person. Indians were the **natives** of North America.

Still the natives didn't understand. Rendova was the name of the base where the PT boats were assigned. Kennedy took a coconut shell and with a knife he wrote the following message on it: "ELEVEN ALIVE NATIVE KNOWS POST AND REEFS NAURU ISLAND KENNEDY." He made signs to the natives for them to take the coconut to Rendova. They finally understood. The navies showed Kennedy and Ross the location of a two-man canoe and then set off in their own canoe toward Rendove.

Then Kennedy and Ross collapsed. Kennedy was very tired and Ross was sick and burned. But they did not stop. Later that night, Kennedy decided that they should take the new canoe and go into the passage where a PT boat might see them. On the way there, they met another storm. The huge waves filled their canoe with water. Both men hung on to one end of the canoe in the storm. Two hours later they were thrown against the coral beaches of the island. They were badly hurt. After reaching the beach, they fell down and slept.

The next morning they saw four big natives standing over them. Were they friendly? They were. One native spoke English very well. He gave Kennedy a letter. The letter was from Lt Evans, Commanding Officer of a New Zealand patrol operating near there.

Saved at last! The men who had remained on the smaller island were very happy to see Lt Kennedy and Ensign Ross return. Kennedy left them and went with the natives to signal the rescue boat. He remained in the bottom of their canoe, covered with branches, so that the enemy could not spot him from the air. They reached the meeting point and heard four shots[12] from the rescue boat. Kennedy answered with four shots. The rescue boat picked up the other men and started towards base. The rescuers, the rescued, and the natives all sang together.

In 1960, John Kennedy became the 35th President of the United States.

Discussion Questions

1. How did bird Island get its name?

2. What was the meaning of the message John Kennedy wrote on the coconut?

3. Why were Kennedy and Ross attempting to swim to another island?

4. Describe what happened to PT-109.

12) **shot**: Sound made by a gun when fired.
Did you hear the **shots** last night?
Yes, I heard five **shots**.

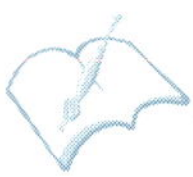

Humor English

A man named Mr. MacGloo was coming home from the market. He spotted a beautiful horse with rippling muscles on his way home. He struck a deal with the owner and at the last minute the owner said "We are a religious family, so we have taught our horse about that. If you want him to gallop say 'Thank God', and if you want to stop you say 'Our Father who art in heaven.'" So the man jumped on the horse and said 'Thank God.' After a few miles there was a cliff ahead. He couldn't think of the verse to make the horse stop. He finally yelled out 2 feet ahead of the cliff, 'Our Father who art in heaven!" The horse stopped and the man got out a handkerchief and said " Whew, Thank God."

Lesson 4.

Why we need forces?

Watching Movie

Top Gun I

Today's Clip

- Hey, Maverick.
- Yeah.
- You hear about Ice?
- What's that?
- He won another one.
- Really?
- Yeah.
- I feel the need.
- The need for speed.
- Gentlemen, this is hop 19. Multiple aircraft, multiple bogies. **Your training is half over.** The competition for the Top Gun trophy

remains tight. **First place**-Iceman. **Second place**-Maverick. Just two points behind.

Vocabulary Checkup

need ⓝ 필요

hop ⓝ 비행

multiple ⓐ 복합적인

Key Expressions

- Your training is half over.
- First place, Iceman. Second place, Maverick. Just two points behind.

Let's Review!

A: Is the Taekwondo competition over?

B: Not yet. **It's only half over.**

A: Who do you think will win?

B: Ummm... I think Jack will probably get **first place**.

Top Gun II

Today's Clip

- Gentlemen, I hate to break up the party before it really gets out of hand... but **there's work to be done.** Some of you have to depart immediately. We have a crisis situation.
- Ice, Slider!
- Yo.
- Hollywood, Wolfman!
- Yes, sir.
- Maverick!
- Sir.
- Maverick, you'll get your RIO. When you get to your ship. And if you don't... **give me a call**. I'll fly with you.
- Sir.

Vocabulary Checkup

break up	방해하다. 끝내다.
gets out of hand	과도해지다. 열기를 더하다.
depart	출발하다
immediately	즉시
crisis situation	비상상황
RIO	부 조종사

Key Expressions

- Gentlemen, I hate to break up the party before it really gets out of hand... but **there's work to be done.**
- When you get to your ship. And if you don't... **give me a call**.

Let's Review!

A: Hello? Is Sergeant Lee in?

B: He is not in at the moment. Who is speaking, please?

A: This is Captain Jackson. Please tell him to **give me a call** as soon as he gets back.

B: Yes, sir.

Pre Reading

1. Working in pairs. Look at the following words and discuss what they mean. Try to give examples for each of them.

Rules of Engagement	**Hostile Act**	**Hostile Intent**
Minimum Force	**Non-Deadly Force**	**Deadly Force**

2. With the whole class. Share your ideas with your colleagues. Give some examples. Do you have examples from your own experience or stories colleagues have told you?

After Reading

3. With a partner prepare a briefing based on information in the article.
4. Change partners, give your briefing to your new partner and listen to your partner giving his briefing. Talk about what was the same and what was different. Comment on how well your partner gave his briefing.(Use Of Force: 영국문화협회)

READING

To shoot or not to shoot?

A soldier is at a checkpoint. A man is shouting at him. A section is on patrol. A crowd gathers and throws rocks at it. A soldier sees a local man lift a rifle and point it at another local person. What do soldiers do in these situations? How do they know how to do the right thing? Well, they have learnt rules to help them. They have learnt the Rules of Engagement, which are the rules that tell a soldier when to use force and when not to use force and how much force to use.

Every conflict and every peace keeping mission has its own Rules of Engagement, and a soldier has a duty to learn the Rules of Engagement for the mission he is on, but there are some things that are common to all operations.

For example, when do peace keepers use force? Well, I must repeat, each PKO will have its own specific rules, but generally force can be used in two circumstances. When there is a hostile act or when there is hostile intent. A hostile act is an actual attack on peace keeping forces, such as throwing bricks at a patrol. Hostile intent is when there is not an actual attack, but there is a strong threat of an attack. A soldier sees a person lifting a rifle and aiming at another person. The soldier believes a person will be killed. The person lifting the rifle has hostile intent and the soldier can use force to stop them from firing.

The first rule is that the soldier should use minimum force. That is, he should only use as much force as is necessary. If a youngman runs at a soldier with a wooden club, that soldier should not shoot the young man. That would be too much force. A trained soldier can take the wooden club off the young man without harming him too much. He would be using minimum force. However, if three young men with knives are running at a soldier and the soldier has no way of escaping, then that soldier could fire his weapon. That too would be minimum force. Minimum force is using the smallest amount of force necessary in the situation. That could be a little or a lot. It depends on the situation.

But soldiers always try to use non-deadly force. That is force that does not use a rifle or other deadly weapon. Examples of non-deadly force are pushing or hitting. Also, a soldier can physically restrain a person, perhaps by grabbing and holding them, or by putting handcuffs on them, or tying their hands and legs in some way. Non-deadly force also includes firing your weapon in the air to give a warning.

But soldiers may sometimes have to use deadly force. That is force which is intended or is likely to cause death or grievous bodily harm, that is serious injury, to a person. Examples of deadly force are shooting at people or using another deadly weapon, such as a knife. But soldiers only use deadly force under conditions of extreme necessity and as a last resort, when all other means have failed or cannot be employed. An example would be if a soldier is being fired on and cannot escape. Then he can return fire and use deadly-force.

Adjusting rifle slings.

On the firing range.

Incident at a Polling Station

On the 12th of June there was an election. It was the first election in the country after many years of violence. "C" company of a peace keeping force were protecting a polling station, that is a place where people vote.

The polling station was in a school on a main square in the town. At about 13:15 hours a large crowd began to gather in the square. Some young men were urg13:15 hocrowd to protest against the election.

Major Roberts, Commanding Officer of "C" company did not like the situation. He decided to negotiate. He sent a translator and a trained negotiator to talk with the young men who were leading the crowd. The translator came back and reported that the leaders would not listen to them. Major Roberts decided to issue a warning. He asked the translator to tell the crowd to disperse, that is leave the square, or he would have to move them by force. The translator picked up a megaphone and asked the crowd to leave.

The crowd then began to pick up stones and throw them at "C" company. Major Roberts ordered a show of force. He deployed his soldiers in two lines facing the crowd and ordered them to take their rifles from their shoulders and have them ready for use in their arms. The crowd still threw stones. Major reports decided to use Non-Deadly force. He ordered his soldiers to move forward as a unit, engage the crowd and push the front row of the crowd back. He also gave permission for his soldiers to hit the people in the crowd on their shoulders or backs, but not on their heads.

A small struggle took place in the square, warning shows were fired and the crowd retreated. Then one young man in the crowd produced a pistol and raised his hand to point it at the soldiers. Rifleman Chambers immediately used deadly force and fired to kill. The young man was hit. The rest of the crowd left the square. The young man who had been shot was still alive. Major Roberts immediately ordered a medical attendant to give the young man first aid and then radioed for a military ambulance. The ambulance arrived and took the young man away.

Major Roberts reported the incidence to his CO.

Lesson 5.

You and Democracy

Watching Movie

A FEW GOOD MEN I

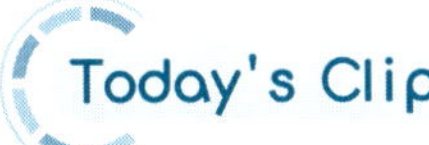

Today's Clip

- Nathan Jessup. Come on in.
- Thank you, sir. Daniel Kaffee. I'm the attorney for Dawson and Downey. This is Lieutenant Commander Joanne Galloway.
- Pleasure meeting you, Commander.
- Observing and evaluating, sir. Lieutenant Sam Weinberg. **He'll be assisting.**
- Sir.
- This is my X.O., Colonel Markinson... and platoon leader Lieutenant Kendrick. **I've asked them to join us.** Sit down, please.

Vocabulary Checkup

attorney	변호사
observe	관찰하다
evaluate	평가하다
assist	도와주다, 보조하다
platoon leader	소대장

Key Expressions

- Lieutenant Sam Weinberg. **He'll be assisting.**
- This is my X.O., Colonel Markinson... and platoon leader Lieutenant Kendrick. **I've asked them to join us.**

Let's Review!

A: This pizza is too big. I don't think two of us can eat it all.

B: That's right. That's why **I've asked Josh to join us.**

A: Have you?

B: Yes. So, don't worry.

A FEW GOOD MEN II

Today's Clip

All right. Everybody **take the night off.**

No, I'm sorry.

I know. Look. We've been working 20 hours a day **for three and half weeks straight**. Just take the night off. Go see your wife, see your daughter. Jo, go do... whatever it is you do when you're not here. What day is tomorrow?

Saturday.

We start at 10:00.

Why do you like them so much?

'Cause they stand on a wall.... and they say, "Nothing's gonna hurt you tonight, not on my watch."

Vocabulary Checkup

take (a day) off	하루 쉬다
whatever	~하는 일이 무엇이든지
wall	벽, 전방
for three and half weeks	3주 반 동안

Key Expressions

- All right. Everybody **take the night off**.
- We've been working 20 hours a day **for three and half weeks straight**. Just take the night off.

Let's Review!

A: What are you doing? Can't you concentrate on it?

B: No, and my brain is not working.

A: Maybe you've been working too much. Why don't you **take the day off**?

B: I wish I could.

Mini Dialog A

W: What's the meaning of democracy?

M: **It means** "government by the people."

Democracy is **more than**[1] a form of government; it's a way of life.

What rights are guaranteed in a democracy?

We **have a right to**[2] **vote** in a secret ballot.

There are **no restrictions on**[3] our property or our jobs.

We **have a right to a fair trial** if we are accused of a crime.

What is the objective of democracy?

To **bring about** a society in which everyone is equal.

In a democracy, we **are entitled to**[4] freedom and justice.

A: Are there **any restrictions on** us at school?

B: Yes. You have **no** right to cheat on tests.

But you **do** have a right to appeal a grade.

1) Democracy is more than... – 민주주의는 …이상의 것입니다.
2) We have a right to... – 우리에게는 …할 권리가 있습니다. to 다음에 명사 또는 동사가 다 쓰인다.
3) There are no restrictions on... – …에 제약이 없다.
4) We are entitled to ~ – 우리는 ~를 누릴 자격이 있습니다.

Mini Dialog B

W : All men are created equal[5]. What does that mean?

M : It **refers to**[6] one's rights and freedoms.

It doesn't **refer to** one's traits or attributes.

How will the class members **deal with** their dilemma?

They'll solve it **through rational discussion.**

They'll deliberate the issue and **cast their ballots.**[7]

A resolution **will be reached**[8] through democratic means.

What is the English teacher **satisfied with**[9]?

She **is satisfied with** Hansu's flawless homework.

She **is satisfied with** Sumi's participation in class.

But she **is not satisfied with** In-ho's lack of effort.

A: Why are you giving me such a severe punishment?

B: It's **for your own benefit.**

You must learn from your mistakes.

5) All men are created equal. - 만인은 평등하게 태어났다.
6) It refers to... - 그것은 …를 가리키는 말입니다.
7) cast a ballot - 투표하다.
8) A resolution will be reached. - 결정이 날 것이다.
9) What is ~ satisfied with? - ~는 무엇에 만족하고 있습니까?

READING

You and democracy

What does the word "democracy" mean?
How does the concept of democracy affect your daily life, and what will it mean to you in the future?

If you look up the word "democracy" in[10] the dictionary, you will find that it comes from a Greek expression, *demos kratos* which literally means "government by the people." *Demos* means the people, and *kratos* means the government. Through the

10) look up a word in ~ – ~에서 단어를 찾아보다 (참고하기 위해 사전 · 지도 등에서 찾을 때).

centuries people have striven to emerge from the shadow of slavery, oppression and persecution into the sunlight of[11] freedom. It has taken hundreds of years and the efforts of millions of people to find[12] ways to make democracy function properly.

The desire for a form of government that would ensure freedom for all citizens[13] has always been strong. As the centuries progressed, nations realized that "government by the people" is the ideal and most egalitarian form of government, because the underlying meaning of this system[14] is government for the benefit of all the people.

Thomas Jefferson said, "all men are created equal." What do these words actually mean? Obviously they do not refer to one's personal traits or attributes;s;s;ssee people of vbuting sizes, utlors and abilities all around ;s;. On the contrbutesthe words "created equal" are interpreted to mean that[15] each individual is born with an equal right to liberty and[16] the pursuit of happiness.

Democracy has a definite objective: to bring about a society in which[17] everyone is entitled to freedom and justice. It is not an entirely flawless system, but it moves slowly but surely towards the ultimate goal of[18] achieving the greatest good for the greatest

11) emerge from... into the sunlight of... – 예속과 압박과 박해의 그늘에서 벗어나 자유의 광명 속으로 나오다.
12) It has taken ~ to find ... – …을 모색하는 데 ~ (시간과 노력) 이 걸렸다.
13) a form of government that would... citizens – 모든 시민에게 자유를 보장해 줄 정부 형태.
14) the underlying meaning of this system – 이 같은 체제의 기본 의미.
this system="government by the people"
15) be interpreted to mean that... – …라는 것을 의미한다고 풀이되다.
16) an equal right to liberty and... – 자유와 행복 추구의 평등한 권리.
17) to bring about a society in which... – 모든 사람이 자유와 정의를 누릴 권리가 있는 사회를 실현하는 일. bring about = accomplish; cause to happen.

number. It is more than merely[19] a system of government. It is a way of life at home and in the community.

Abraham Lincoln delivering the Gettysburg Address

What kind of freedoms and rights do you have in a democracy? You have freedom of speech and assembly. You have the freedom to worship in any way you wish[20]. You have freedom of the press, which enables you to both read, print and distribute whatever you like, and to[21] criticize the government when you disagree with its policies. There are no restrictions on the property you may own, the business you may establish, or[22] the profession you may enter.

You have the right to a fair trial[23] in a court of law if you are accused of a crime[24]. You are considered innocent[25] until you are proven guilty[26]. If you feel that you have been wrongly convicted, or that your punishment is too severe, you may appeal the decision[27] and ask the courts to reconsider[28].

18) it moves... towards the ultimate goal of... - 민주주의(it)는 최대 다수의 최대 행복을 이룬다는 궁극적 목표를 향해 …나아간다.
19) It is more than merely... - 민주주의는 단순히 …만은 아니다(…이상의 것이다.) it=democracy. merely=simply, only.
20) in any way you wish - 어떤 방식이든 원하는 대로.
21) enable you to ~ and to ... - 우리로 하여금 ~ 하고 … 할 수 있게 해 주는.
22) There are no restrictions on A, B, or C - A, B 또는 C에 대한 제약이 없다.
23) the right to a fair trial - 공정한 재판을 받을 권리.
24) be accused of a crime - 어떤 죄로 기소되다.
25) be considered innocent - 무죄로 간주되다.
26) be proven guilty - 유죄로 입증되다.
27) appeal the decision - 판결을 항소하다.
28) ask the courts to reconsider - 법원에 재심을 청구하다.

Then there is the greatest freedom of all-the right to vote in a secret ballot for the candidate who you think is most suitable[29]. You are at liberty to discuss any public issues that concern you, and to participate in[30] the decision-making process. You may even run for public office[31] yourself, if you choose to do so.

The government leaders in a democracy are elected by popular vote. They are brought to power[32] by agreement of the people, and this same power can also be withdrawn by the electorate. If the voters are dissatisfied with a representative, they have a peaceful and legal means of removing him from office at the next election.

In a democracy there are problems to be dealt with and dilemmas to be[33] resolved at all times. Rational discussion, reason, and casting one's ballot are the democratic means used to[34] reach a peaceful resolution. After all sides of the issue have been presented, deliberated and weighed carefully, a final decision is made by a vote of the majority. But the opinion of the minority is also respected. Reason and free discussion are far more powerful than all the guns and bombs in the world.

Democracy does not belong to any particular racial group. It

29) ~who is most suitable – 가장 적합한 ~.
30) be at liberty to discuss ~, and to participate in... – ~을 자유롭게 토론할 수 있고 …에 자유롭게 참여 할 수 있다.
31) run for public office – 공직에 출마하다.
32) be brought to power – 권력을 잡게 되다.
33) problems to be... and dilemmas to be... – 처리해야 하는 문제와 결정지어야 하는 곤란한 일들.
34) the democratic means used to... – …하기 위해 사용되는 민주적 수단.

belongs to all the people who recognize that[35] every person should be respected as a human being. Today, no country can remain isolated. Nations as well as individuals need each other, and everyone must co-operate to bring about and preserve peace[36]. This can be accomplished only when all people everywhere are granted[37] equal rights and freedoms.

The most important guardian of democracy is the understanding of democracy itself. Democracy will never be secure unless you understand it, believe in it and live by it. You can see what a great responsibility each person has[38] in the welfare of the country. You may be only one individual, but your participation is crucial. The leadership of the country will depend upon today's youth. You will have to train yourself for good citizenship, and you will have to carry the torch of liberty and justice for all.

35) all the people who recognize that... – …이라는 것을 인식하고 있는 모든 사람.
36) co-operate to bring about and preserve peace – 평화를 이룩하고 보존하기 위해 협력하다.
37) be granted... – …을 부여받다.
38) what a great responsibility each person has – 개개인에게 얼마나 큰 책임이 있는지를 (= how great a responsibility...).

Talking about the Right to Vote

A: Where are you going, Mom?

B: I'm going out to vote in the election.

A: Why do you bother[39]?
Does your vote count[40]?

B: Of course. Everyone's vote is important. We have a say in choosing[41] our leader.

A: That's because Korea is a democracy, isn't it?

B: Yes. The right to vote is our greatest freedom.
You'll be able to vote when you're twenty.

A: By the way[42], Mom, who are you going to vote for[43]?

B: It's a secret! That's why[44] we have a secret ballot[45].

---

39) Why do you bother? – 성가시지 않으세요?
40) Does your vote count? – 어머니 표가 중요해요?
41) We have a say in ~ing... – 우린 …를 ~할 결정권을 갖고 있단다.
42) by the way – 그건 그렇고, 말이 나온 김에.
43) Who are you going to vote for? – 누구를 찍으실 거예요?
44) That's why... – 그게 바로 …하는 이유란다.
45) a secret ballot – 비밀 투표.

Reading Comprehension

A. Wh Questions.

1. Which language does the word "democracy" come from?
2. What does "democracy" really mean?
3. What is the main goal of democracy?
4. What rights do we have in a democratic society?
5. How are government leaders brought to power?
6. Who is able to vote, or run for office?
7. How are problems solved and decisions made?
8. Whom does democracy belong to?
9. Why is your participation important in a democratic system?

B. Select the unsuitable item from the list below to complete the main text.

1. Democracy is often considered to be an ideal form of government.
   (a) Democracy is a very new and modern concept.
   (b) Democracy is government by the people.
   (c) Reason is more powerful than weapons.
   (d) Everyone is entitled to freedom and justice in a democracy.

2. We all have certain rights and freedoms in a democratic society.
   (a) We may join any political groups we choose.
   (b) We can choose our own government representatives.
   (c) We are permitted to write anything we want.
   (d) If we are convicted of a crime, there is nothing we can do.

A famous Korean poem translated in English

I Never Knew Before

Kim So-wol(1865-1939)

That the moon rides the night[46], spring or fall,
I never knew before[47].

That I would miss[48] you with such pain[49],
I never knew before.

That one could reflect on the moon[50] in all its brightness[51],
I never knew before.

And that the moon is sadness itself,
I never knew before.

46) That the moon rides the night – 밤이면 달이 뜨는 것(을).
47) I never knew before – 예전엔 미처 몰랐어요.
48) miss~ – ~를 그리워하다.
49) with such pain – 그처럼 아프게.
50) reflect on the moon – 달에 반사되다, 비추이다.
51) in all its brightness – 한껏 밝게; 환하게.

Lesson 6.

The Korean War

Watching Movie

BAND OF BROTHERS I

Today's Clip

- Colonel Sink **has seen fit** to promote you.
  As 1st Lieutenant, **you'll serve as my executive officer.**
  Congratulations.
- Thank you, sir.
- And, as a test of your organizational skills and command potential, I am designating you... Mess officer for 14 days. Report to the mess kitchen at 0515 hours. Company breakfast to be served at 0600.
- Yes, sir.

Vocabulary Checkup

fit	맞다 적합하다
executive officer	선임 장교, 부지휘관
organizational skills	조직력
command potential	통솔력
designate	임명하다
mess officer	(군대) 식당, 취사장

Key Expressions

- Colonel Sink **has seen fit** to promote you.
- As 1st Lieutenant, **you'll serve as my executive officer.** Congratulations.

Let's Review!

A: I know you are not going to like this, but the Testing Department **has seen fit** to change the format of the test. It will no longer be multiple choice, but will be fill in the blank.

B: What?! I can handle a multiple choice test, but fill in the blanks?!

A: I know, I know. Just relax. We'll take a few practice tests to help prepare you.

BAND OF BROTHERS II

Today's Clip

- How many men have Dog company got assembled?
- **A handful,** maybe 20.
- You are the only officer that made it?
- So far. **Still waiting for** orders. You got some cigarettes?
- Here. Hey, keep the pack.
- Yes, sir.

Vocabulary Checkup

get assembled	모이다, 집합하다
handful	한줌, 소수
so far	지금까지
pack	상자

Key Expressions

- How many men have Dog company got assembled?
- **A handful,** maybe 20.
- You are the only officer that made it?
- So far. **Still waiting for** orders.

Let's Review!

A: I'm **still waiting for** you to tell me about how you got that black and blue eye.

B: Well, I was attacked in the parking lot, while trying to get inside my car.

A: What? Who was it?

B: It was just **a handful of** middle school girls...

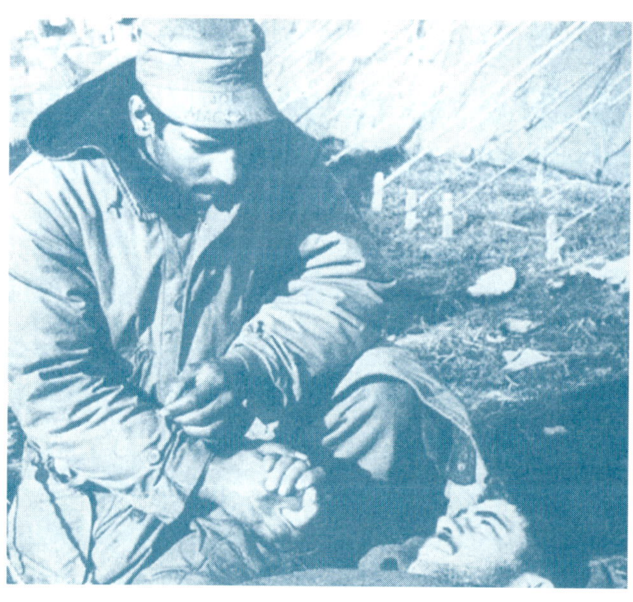

BAND OF BROTHERS III

Today's Clip

- Remember, boys.. Give me three days and three nights of hard fighting and **you will be relieved.**
- Lieutenant, I'll take point.
- Corporal Hoobler will be lead scout.
- Blithe, **glad you could join us.**
- Thank you, sir.
- 1st platoon, fall in behind Fox Company. You people from 2nd and 3rd platoon, follow us.
- Let's shake a leg.

Vocabulary Checkup

relieved	해방되다
point	선봉, 첨병
scout	정찰병
fall in	정렬하다
platoon	(보병, 공병의) 소대
shake a leg	바쁘게 걷다, 서두르다

Key Expressions

- Remember, boys.. Give me three days and three nights of hard fighting and **you will be relieved.**

- Blithe, **glad you could** join us.
- Thank you, sir.

Let's Review!

A: **You will be relieved** in four hours.

B: Four hours! I can't stay that long - my wife is all alone at home.

A: Look! Bill's here! Bill, **glad you could join us.**

B: Good. Bill can take my place so I can go home early.

Mini Dialog A

W: Whose statue is that standing over there?

M: It's General Douglas MacArthur's.

He worked out the details of the Inchon landing.

He helped us to turn defeat into victory.

How did the United Nations respond to the invasion?

It **called** the Security Council delegates **into** a special session.

They decided to **furnish** the R. O. K. **with**[1] assistance.

More troops **were required to** slow down the Red advance[2].

The South Koreans **were lacking in** tanks and artillery.

They **were outnumbered** three to one[3].

The North Koreans were **under cover of** an air umbrella[4].

A: Why don't we go this weekend, Yu-mi?

B: Now is the best time to visit Mt. sorak.

---

1) furnish Korea with... - 한국에 원조를 제공하다.
2) slow down the Red advance - 붉은 군대의 전진을 지체시키다.
3) They were outnumbered three to one. - 그들은 3대 1로 숫자가 열세였다.
4) under cover of an air umbrella - 공군의 엄호를 받으며.

Mini Dialog B

W: What did your friends do yesterday?

M: Tom went to Kate's in the hope that[5] she would be there.
Bill stayed in bed in the hope that he would get better.

What were the events of[6] the Inchon landing?
The marines landed at Wolmido and **secured a beach-head.**
General MacArthur **set foot on**[7] the island.
General Walker **slowed** the enemy **to a walk.**
The North Koreans were captured **by the thousands**[8].

What happened during the storm?
Several farms **were cut off from** supplies[9].
Some cities and towns **were cut off from** electricity.

A: Did you witness[10] the car accident?

B: Yes. I was just in time to witness the accident.
There was a tower of smoke rising from it.
The parts of the car were scattered everywhere.

---

5) in the hope that... – …하기를 기대하며.
6) What were the events of...? – …은 어떤 사건이었습니까?
7) set foot on – ~에 상륙하다.
8) by the thousands – 수천 명씩
9) be cut off from supplies – 보급품이 끊기다.
10) Did you witness...? – …을 목격했습니까?

READING

Douglas MacArthur

The following is General MacArthur's account of the Korean War[11] — how it started, how Seoul was lost and restored, and how he fought to turn defeat into victory[12].

It was early morning, Sun-day, June 25, 1950, when the telephone rang at the American Embassy in Tokyo. It was the duty officer at headquarters. "General," he said, "we have just received a briefing from Seoul, advising[13] that the North Koreans struck south across[14] the 38th Parallel at four o'clock this morning."

I had a feeling of nightmare[15]. The South Koreans had four divisions along the 38th Parallel. They had been well trained, and were courageous and patriotic. But they had only light weapons, no air force or naval force, and were lacking in tanks, artillery and other essentials. The Communists had masked their preparations for attack. Along the border they had only a lightly armed force, but behind this first line of offense was a powerful army[16], fully

11) MacArthur's account of the Korean War - 맥아더의 한국 전쟁 이야기.
12) turn defeat into victory - 패배를 승리로 바꾸다.
13) have just received a briefing, advising... - …을 알리는 보고를 방금 받았다.
14) struck south across ~ - ~을 넘어 남쪽을 공격했다.
15) I had a feeling of nightmare. - 나는 악몽을 꾸는 기분이었다.

equipped with heavy weapons, including the latest model of Soviet tanks.

Now was the time to[17] recognize what the history of the world has taught from the beginning of time: that timidity brings conflict, and courage often prevents it. The Secretary-General of the UN summoned the Security Council delegates into a special session[18]. The Security Council concluded that urgent measures were required[19] to restore peace, and that the Republic of Korea should be furnished with any necessary assistance. Thus the United States went to war against Communism in Asia[20].

16) behind this first line of offense was a powerful army - 이 1차 공격선 뒤에는 강력한 군대가 있었다. 부사가 앞에 온 문장.
17) Now was the time to.... - 이제 …할 시간이었다.
18) summon ~ into a special session - ~을 특별 회의에 소집하다.
19) urgent measures were required - 긴급 조치가 필요했다.
20) went to war against Communism in Asia - 아시아에서 공산주의와의 전쟁을 치르게 되었다.

On June 29, I flew to Korea to see the disastrous situation for myself. The capital city of Seoul was under heavy attack[21], and the Korean government had moved to temporary headquarters at Taejon.

My plane landed at Suwon, 20 miles south of Seoul[22]. I took a jeep and headed north towards the Han River under constant bombing. We reached the banks of the Han just in time to witness[23] the last action to defend its bridges. Seoul was already in enemy hands, and across the river I could see towers of smoke[24] rising from the ruins. It was a tragic scene of agony and destruction. The troops were withdrawing, and here and there I could see the bright red crosses of ambulances.

The sky was filled with the shrieking sound of shells[25], and everywhere was the smell of death[26]. The refugees painfully walked south, carrying their belongings on their backs and leading their terrified children. For an hour I watched the pitiful

21) ...under heavy attack – 심한 공격을 받고 있었다.
22) 20 miles south of Seoul – 서울에서 남쪽으로 20마일 떨어진.
23) reached ~ just in time to witness... – ~에 때맞춰 도착하여 …을 목도하게 되었다.
24) towers of smoke – 높이 솟은 연기 기둥들.
25) the shrieking sound of shells – 포탄의 찢어지는 듯 한 굉음.
26) and everywhere was the smell of death – 그리고 사방이 죽음의 냄새로 차 있었다.

evidence of the disaster, and there on the banks of the Han I made up my mind.

I was convinced that[27] the South Koreans could not stop the enemy's sweeping rush south. I would throw my occupation soldiers in Japan into Korea. We would be outnumbered three to one; it would be desperate, but it was my only chance. Gradually the idea of the Inchon landing operation began to take shape[28]-a counterattack that could turn defeat into victory.

The North Koreans had an estimated strength of 200,000 men, spearheaded by approximately 200 Soviet tanks, all under cover of an air umbrella. The immediate priority[29] was to slow down the Red advance before it enveloped all of Korea. I had to commit my forces as rapidly as I could to the front.

In August the South Koreans were reorganized, and five R. O. K. divisions were with General Walker. They withdrew from Taejon and positioned themselves along[30] the Naktong River. Fresh divisions and tank units from North Korea arrived constantly, but General Walker, with great skill and courage, slowed the enemy to a walk[31].

I was finally ready to bring my plan into fruition[32]. The target I selected was Inchon. The target date, because of the

27) be convinced that ~ - ~을 확신하다.
28) take shape - 모양이 갖춰지다, 구체화되다.
29) The immediate priority - 최우선적으로 할 일.
30) positioned themselves along... - 낙동강을 따라 포진했다.
31) slow the enemy to a walk - 적을 걷는 속도로 지체 시키다.
32) bring one's plan into fruition - 계획을 실현시키다.

great tides at Inchon, had to be the middle of September. A week before the target date, all the details of the master plan[33] had been worked out.

On September 12 my ship, the Mount McKinley, departed from Japan. I had made many previous landings before, but this was the most complicated amphibious operation[34] I had ever attempted. By the second day at sea, the tension on board was[35] approaching its climax. Soon, 40,000 men would act boldly in the hope that 100,000 others along the Naktong River would not die. I alone was responsible, and if I failed, the dreaded results would rest on judgment day against my soul[36].

A sudden thunder woke me that night. Our guns had opened up on[37] Wolmido. This harbor island was rocking under the

33) all the details of the master plan... – 주 작전 계획의 세부 사항들이 모두 작성되었다.
34) amphibious operation – (육·해·공) 합동 작전.
35) the tension on board was... – 배 안의 긴장이 최고조에 달하고 있었다.
36) judgment day against one's soul – 영혼의 최후 심판 날.
37) open up on ~ – ~에 포격을 개시하다.

bombardment of naval guns and aerial bombs, and the first landing craft were going in. At 8 a. m. I was informed that the first wave of Marines[38] had landed and secured a beachhead.

An hour later, Wolmido had been fully secured. Late that afternoon I set foot on the island. I directed the prompt capture of[39] Seoul and the advance towards the south[40]. This would place the enemy between[41] two giant forces-one from the north, one from the south.

One column headed for Seoul, and the other column moved towards Suwon to capture the air base. Kimpo Airfield, the largest in Korea, was captured. I directed General Walker to attack, and he promptly crossed the Naktong River. The enemy resisted, but was cut off from supplies, command and communications. They could not help retreating[42].

North Korean prisoners surrendered or were captured by the thousands. Within a month, the total Red captives rose to 130,000. The enemy fought desperately for Seoul, but by September 28 it had been liberated. The government of Korea was once again established in Seoul. At the Capitol Building the ceremony of handing back the capital[43] began. And, when the ceremony was over, the people of Seoul lined the streets[44] and waved their little paper flags. I returned to Tokyo.

38) the first wave of Marines - 해병 선발대.
39) direct the prompt capture of - ~의 신속한 탈환을 지시하다.
40) the advance towards the south - 남쪽을 향한 전진.
41) place the enemy between... - 적을 막강한 두 군대 사이에 몰아넣다.
42) They could not help retreating. - 그들은 후퇴할 수밖에 없었다.
43) the ceremony of handing back the capital - 수도탈환 기념식.
44) line the street - 연도에 늘어서다.

Talking about a Trip to Panmunjom

A: Good morning. We'd like to[45] visit Panmunjom.

B: Well, you can only go there with a tour group.

A: What will we be able to see there?

B: You'll see the Freedom Bridge, and the Conference Room.

A: Are there any special rules we should follow?

B: Yes. Don't talk to[46] any North Korean soldiers.

A: I'd like to make a reservation for two. How much are the tickets?

B: They're 45,000 won each. You'll have a fascinating[47] experience.

45) We'd like to... - 우리는 …하고 싶습니다.
46) Don't talk to... - …와 이야기하지 마십시오. talk 다음에 with 나 to가 온다.
47) fascinating - a. 매혹적인.

Reading Comprehension

A. Wh Questions.

1. <u>When</u> did the Korean War break out?
2. <u>How</u> was the South Korean army equipped at the time?
3. <u>Where</u> did the government of the R. O. K. move to?
4. <u>How</u> many North Korean prisoners were captured?

B. True / False Questions.

1. The South Koreans had been expecting an invasion. T / F
2. The North Korean army easily won control of Seoul. T / F
3. MacArthur had enough soldiers to outnumber the Communists. T / F
4. The Inch'ŏn operation turned out to be a success. T / F

C. Select the unsuitable item from the list below to complete the main text.

1. Why was the North Korean invasion so successful at the beginning?
   (a) The North Koreans gave no warning of the attack.
   (b) The Communist forces had the latest Russian tanks.
   (c) The United Nations did not want to help South Korea.
   (d) South Korea did nat have many troops or weapons.

2. Why was the Inchon landing to be a difficult operation?
   (a) General MacArthur had never made a landing before.
   (b) MacArthur's soldiers would be outnumbered by the enemy.
   (c) The details of the plan were very complicated.
   (d) The tides around Inchon were sometimes a problem.

Humor English

Who Stole a kiss?

Four people were riding in a train coach. A woman and her beautiful nineteen-year-old daughter were on one side, and facing them were the army general and his escort, an army private. The train entered a tunnel, and the cabin became dark. A kiss was heard, followed by a slap.

The mother thought. "That young man stole a kiss from my daughter, and she rightfully slapped him."

The daughter thought, "That young man tried to kiss me and kissed my mother by mistake and got slapped."

The general thought, "That young man stole a kiss, and I got slapped by mistake."

The private thought, "I'm pretty smart. I kissed the back of my hand and got to hit the general."

누가 키스를 몰래 했나?

네 사람이 기차를 타고 가고 있었다. 한 여인과 그녀의 열아홉 살 아름다운 딸이 한 쪽에 있었고, 육군 장군과 그의 호위병인 육군 사병이 그들을 마주보고 있었다.

기차가 터널을 들어서자 객실이 어두워졌다. 키스 소리가 들리고 뒤이어 뺨 때리는 소리가 들렸다.

"저 젊은 놈이 내 딸한테 몰래 키스해서 딸애가 정당하게 그를 때린 거야"라고 어머니는 생각했다.

"저 젊은 놈이 나한테 키스하려다 실수로 어머니에게 키스를 해서 얻어 맞은거야"라고 딸은 생각했다.

"저 젊은 놈이 몰래 키스했는데 내가 잘못 얻어맞았어."라고 장군은 생각했다.

"난 꽤 똑똑하단 말이야. 내 손등에 키스하고 장군을 때리게 되었으니"라고 사병은 생각했다.

Lesson 7.

The Gentleman in White

Watching Movie

BAND OF BROTHERS IV

Today's Clip

- Couple of announcements, men. First, listen up. First, the training exercise scheduled for 2200 **has been cancelled.**
- Oh, yeah..
- Secondly.. all passes are hereby revoked. **We're heading back to** France. So, pack up all your gear. We will not be returning to England, boys. Anyone who has not made out a will, go to the supply office. The trucks depart for Membury at 0700. As you were.

Vocabulary Checkup

scheduled for	~로 예정된
pass	통로
revoked	막히다
pack up	(짐을) 꾸리다
gear	군장
make out a will	유언장을 쓰다
supply office	보급소
As you were.	바로. 이상.

Key Expressions

- Remember, boys.. Give me three days and three nights of hard fighting and **you will be relieved.**
- Blithe, **glad you could join us.**
- Thank you, sir.

Let's Review!

A: **You will be relieved** in four hours.

B: Four hours! I can't stay that long - my wife is all alone at home.

A: Look! Bill's here! Bill, **glad you could join us.**

B: Good. Bill can take my place so I can go home early.

BAND OF BROTHERS V

Today's Clip

- What do you think you'll do after this?
- Get some breakfast.
- No, after, after.
- Well, it's funny you should mention it. 'cause I had a meeting with Colonel Sink.
- Really?
- Yeah, he **discussed the possibility of** staying.
- In the army?
- Yeah, as a career.
- What'd you say?
- **I said I'd think about it.**

Vocabulary Checkup

mention	언급하다
'cause = because	왜냐하면
possibility	가능성
career	직업

Key Expressions

- Yeah, he **discussed the possibility** of staying.
- What'd you say?
- **I said I'd think about it.**

Let's Review!

A: What do you want to do after graduation?

B: My father and **I discussed the possibility** of being a teacher. What do you think?

A: It's a stable and popular job these days. What did you tell your Dad?

B: **I said I'd think about it**. To be honest, I'm not so sure....

Mini Dialog

A Swiss made a suggestion.

A: Where did the Red Cross emblem come from?

B: It was modelled after the Swiss flag.

A: How come?

B: A Swiss made a suggestion and the Swiss people played a major role in establishing the organization.

The Red Cross came into being through the efforts of Dunant.

A wounded enemy is an enemy no longer.

A: One person can change the world for the better.

B: I agree. The Red Cross came into being through the dedicated efforts of Henri Dunant.

A: Yes, he was a great man. He realized that wounded soldiers needed better medical treatment.

B: That's right. To him a wounded enemy was an enemy no longer.

READING

The Gentleman in White

Henri Dunant

Most of us today have heard about the Red Cross and its great services throughout the world. This humanitarian organization was established in 1863 through the dedicated efforts of Henri Dunant. The following piece gives a brief summary of how he was able to establish the Red Cross against great adversity.

No one paid much attention to[1)] a young Swiss gentleman who was taking a sojourn[2)] in Italy in 1859. He was reserved[3)] and polite. In both dress and manner he had the aristocratic air[4)] of one born to wealth and position[5)]. But something that happened in the midst of[6)] his tour changed young Henri Dunant into a fervent reformer determined to right a great wrong[7)].

Italy and France were at war with Austria. Henri Dunant came upon[8)] one of their bloody battlefields. It was Solferino at the

1) pay attention to : attend to – 주의를 기울이다.
2) sojourn : a stay in a place other than one's home for a time – (타향·타국에서의) 체재, 체류.
3) reserved : that does not like to talk about himself or to make this feelings known – 말수가 적은, 언행을 삼가는, 쌀쌀한.
4) air : the general character or appearance of something; appearance of, or feeling caused by, a person or place – 풍모, 분위기.
5) born to wealth and position – 재산과 지위를 가진 사람에게 태어난.
6) in the midst of : in the middle of; during – ~의 가운데, ~하는 중에.
7) right a great wrong – 커다란 잘못을 바로잡다.
8) come upon : meet by chance ; come across – 우연히 마주치다, 우연히 발견하다.

close of[9)] a day of fighting. For the first time in his life, Dunant saw how cruel and heartless war could be. All round him lay suffering men, groaning in agony[10)], untended[11)] and left to die where they fell.

Dunant helped to save hundreds of lives with his assistants.

Sickened at heart[12)] by what he had seen, Henri Dunant went to work at once. Helped by several peasant women, he formed an ambulance[13)] service. He set up headquarters[14)] in a little church not far away, and he made use of anyone who came along. In fact he himself help to bind up the wounds of Frenchmen, Italians, and Austrians alike[15)]. Someone asked Dunant why he cared for Italy's enemies. His answer showed a spirit of humanity that was uncommon in those days. "We are all brothers[16)]. A wounded enemy is an enemy no longer."

---

9) at the close of : at the end of – ~의 마지막에.
10) agony : very great pain – 고통
11) untended : not being tended to ; not being attended to – 돌보는 이 없는, 간호 받지 못한.
12) Sickened at heart = As he was sickened at heart
13) ambulance – ① 앰뷸런스, 구급차 ② 야전 병원.
14) headquarters – 본부, 사령부(단수 · 복수로 취급), a headquarters라고 표현함.
15) alike : the same way; equally – 같은 모양으로, 마찬가지로 (A and B alike 형으로 써서 both A and B의 의미를 나타낸다.

With his hearty band[17] of assistants, Dunant helped to save hundreds of lives. Water and medicine were brought. Pallets[18] of straw were smoothed so that the wounded could lie in comfort[19]. Those who despaired were given hope. The last prayers[20] of the church were said for the dying[21]. The "gentleman in white[22]" was looked upon by thousands of wounded men as[23] little short of an angel[24].

Henri Dunant was too modest to dwell upon his work in helping the sufferers. But he could not forget what had happened at Solferino. He could not erase the memory of the needless suffering on the battlefield. The soldiers had been left to suffer, some of them to die, because no group of people had been ready to[25] look after the wounded. The more he thought of it, the more he felt that something must be done.

He hit upon[26] the idea of a great organization. It should be planned to take in[27] many nations. It should have everything

16) We are all brothers - all 이 주어의 뒤에서 쓰일 때는 대개 빈도 부사의 위치라고 알아 두면 좋다.
17) band : a group of people formed for some common purpose and often with a leader - 무리, 단체
18) pallet : hard narrow uncomfortable bed - 간이침대.
19) in comfort : comfortably - 편안하게.
20) prayer - 기도의 말
21) the dying - 죽어가고 있는 사람들.
22) the gentleman in white - 보통 간호원을 "an angel in white (백의 천사)"라고 부르는 데서 유추한 표현.
23) look upon A as B - regard A as B - A를 B라고 여기다, 간주하다.
24) little short of : almost; little less than - ~와 거의 같은
25) be ready to : be prepared to; be willing to - ~할 준비가 되어 있다, 기꺼이 ~ 하다.
26) hit (up) on : find by lucky chance or have a good idea about - (생각을) 떠 올리다, (좋은 수를) 발견하다.
27) take in : permit to enter - 끌어들이다.

needed to relieve human suffering, especially in time of war. It would do its work with the approval of all countries.

The headquarters of the Red Cross

Bit by bit[28)] the plan crystalized in his mind. Each detail grew clearer. All nations would support and share in[29)] such a great work of humanity. It must operate under a sign that all would know: its symbol would be a red cross against a white background.

One man alone could hardly hope to make such a grand vision come to pass[30)]. But Henry Dunant resolved to do what he could. First of all he must reach the public. It must see the need for this great organization. To make the need clear, he wrote a truthful account of what he had witnessed at Solferino. At the end of his vivid account he made a suggestion : an international relief society should be set up to care for the wounded.

Dunant's book, *A Memory of solferino,* shocked and startled its readers. They were appalled[31)] by the bloodshed[32)] he described.

28) bit by bit : by degrees ; little by little, a bit at a time, a little at a time – 조금씩 조금씩, 점점
29) share in : to have a share in ; take part in – ~에 끼다, 참여하다.
30) come to pass : happen; come about – 일어나다, 발생하다.
31) appal : shock deeply ; fill with fear – 경악시키다, 오싹하게 하다, 몸서리치게

They were shocked by the lack of care for the wounded[33]. And they were startled to have the author prove that these horrors he described did not have to be. All that was needed was[34] a relief society ready to give first aid[35] on the battlefield.

The Red Cross activities

One of Dunant's own countrymen was the first to act. Gastav Meunier, a wealthy lawyer in Geneva, had been deeply stirred[36] by the book. He chose Dunant and four others to form the Committee of Five. It was to look into[37] the idea of an international society.

After some intensive[38] study, the committee invited all the nations of Europe to a meeting in Geneva. The conference, held on October 26, 1863, was a moderate[39] success. Thirty-six

하다.
32) bloodshed : killing in fighting - 살육
33) lack of care for the wounded : - 부상병들에 대한 구호의 결핍 (care 는 명사)
34) All that was needed was... - 필요한 것은 오직 … 이었다.
35) first aid - 응급 치료
36) stir : to excite the feellings of ; move - 감동시키다, 흥분시키다.
37) look into : to examine the meaning or cause of - 조사하다, 면밀히 검토하다.
38) intensive : which gives a lot of attention or action to a small amount of something - 집중적인

delegates[40] representing fourteen nations were present. Dunant's suggestions were approved. It was decided that a relief society should be formed in each country.

The international agreement called the Geneva Convention[41] was made the following year. The nations of Europe approved the new organization. For the first time in history, they accepted wounded men as neutral[42] : a wounded soldier would no longer be looked on as an enemy. The nations that signed the Geneva Convention chose the red cross as[43] the symbol for the international organization. At first the new society was given various names in different countries. Later it was known across the globe[44] as the Red Cross Society.

Until his death in 1910, at the age of eighty-two, Henri Dunant was always ready to work for a noble cause[45]. And when the Nobel prize for peace[46] was awarded[47] for the first time, in 1901, the honor went to Dunant. No man has deserved[48] it more than the founder of the Red Cross.

---

39) moderate : of middle degree; neither large nor small, high nor low, fast nor slow - 어느 정도의, 평범한, 수수한,(중간정도나 그 이하의 경우에 호의적으로 쓰는 용어)
40) delegate : a person acting for one or more others, such as a representative to a meeting or an organization - 대리인, 대표자
41) Convention : a formal agreement - 협정, 조약
42) neutral : being or belonging to a country which is not fighting or helping either side in a war - 중립국의
43) choose A as B = choose A to be B - A를 B로 선출하다, 선택하다.
44) across the globe - 전 세계에 = all over the world
45) cause - 대의명분, 주장, 신조 = a principle or movement strongly defended or supported
46) the Nobel prize for peace - 노벨 평화상
47) award : to give - ~에게 ~을 수여하다.
48) deserve - ~을 받을 만한 자격이 있다. ~을 받을 만하다.

Reading Comprehension

A. Wh Questions

1. What did Dunant see at Solferino?
2. What did Dunant form?
3. Why did Dunant care for Italy's enemies?
4. What plan did Dunant have in mind?
5. Who helped Dunant form the Red Cross Society?

B. True / False Questions

If the statement is false, rewrite it to be true.

1. Henri Dunant was known as the "gentleman in black."
2. Dunant hoped to make an organization to relieve human suffering.
3. The readers of Dunant's book were not surprised.

C. Summary Question

In 1859 in Italy Henri Dunant saw how heartless ________ could be. All around him lay suffering men, untended and left to ________ where they fell. Helped by several and left to ______ where they fell. Helped by several peasant women, Dunant formed an ______. He helped to save many lives, and was looked upon by hundreds of wounded men as little short of an ________. He wanted to form an international organization to relieve ________. One of Dunant's own countrymen, ________ formed a Committee of Five. The Geneva Convention chose the red cross as the ________ of the Red Cross Society. Dunant won the ________ for peace in 1901.

An English poem

When June is Come

- Robert Bridges(1844-1930) -

When June is come, then all the day,
I'll sit with my love in the scented hay,
And watch the sunshot palaces high
That the white clouds build in the breezy sky.

She singth, and I do make her a song,
And read sweet poems whole day long;
Unseen as we lie in our haybuilt home,
O, life is delight when June is come.

Lesson 8.

On a Military Base

Watching Movie

Saving Private Ryan I

Today's Clip

- **You came all the way out here to** tell me that?
- You're going home. Our orders are to bring you back.
- Bring me back?
- Corporal Anderson, **I don't mean to** leave you even more short-handed, but **orders are orders**.
- Any communications about when you will be relieved up here?
- There's no way to tell. We have no idea what's happening south of us.
- I have my orders, too, sir. They don't include me abandoning my post.
- I understand that, but this changes things.

- I don't see that it does, sir.
- The Chief of Staff for the US Army says it does.

Vocabulary Checkup

short-handed	인원부족의
communications	소식, 정보
to be relieve	구조되다
abandon	포기하다

Key Expressions

- You came all the way out here to tell me that?
- Corporal Anderson, **I don't mean to** leave you more short-handed, but orders are orders.

Let's Review!

A: **I don't mean to** keep complaining, but aren't those documents ready yet?

B: Well, they were more difficult than I thought. Can you wait a second? They will be right out.

A: Okay.

Saving Private Ryan II

Today's Clip

- Might as well be spit wads if they roll on us with tanks.
- Which they're sure to do.
- What are you thinking, sir?
- I'm thinking I'm going to whip around the flanks. Unless we can draw them up this main road here between these buildings, where all the rubble makes a bit of a bottleneck...
- Disable him?
- If we can. Make that tank 60-ton roadblock. If we do that, we've got a fighting chance at her flanks.
- That's going to split them up, don't let them mass.

Vocabulary Checkup

spit wads	작은 솜들
flank	측면
rubble	파편
bottleneck	병목, 좁은 길
a fighting chance	성공 가능성

Key Expressions

- What are you thinking, sir?
- That's going to split them up, don't let them mass.

Let's Review!

A: Are you looking forward to the Taekwondo competition? It will be held this Friday.

B: Good! Last year's winner is still injured, so with him out of the way, I have **a fighting chance** this year!

READING

1. A Military Installation

The installation where we are stationed is typical of many installations. The facilities cover an area of one hundred square miles. In addition to the facilities that are usually found on any military installation, there are special buildings for classroom instruction. There are also library facilities, repair and maintenance areas for equipment, and hospital facilities.

Officers' Club

Dining Hall

Housing includes barracks for the servicemen, complete with recreation and study facilities. Both apartments and houses are available for married officers whose families are with them. These military quarters are equipped with modern electrical appliances. There are also playgrounds for the small children. A commissary is operated for servicemen and their families.

Recreation and sports are part of the life of the men at the installation. Baseball, basketball, football, volleyball, and soccer teams are organized. There are tennis courts, indoor and outdoor swimming pools, and recreation rooms containing equipment for table tennis.

Movies are shown several times a week in the theaters. There are frequent visits by entertainers who present programs of music, comedy, or drama. At the installation there are chaplains of the major religious faiths. There are also special counseling services for those who seek advice on career or personal problems.

All in all, the installation is a complete community in itself. It provides most of the activities of American life. In addition, the installation provides training necessary for the officers and men assigned to it.

2. Military Customs and Courtesies

As a military student in America, you will be with us for an extended period of time. We know you have some questions concerning our military customs and courtesies. The purpose of this reading is to inform you of more important customs and courtesies of our Armed services. You will notice that many of these customs are similar to those practiced in the military of your own country. We will not discuss all of the customs, only those that we feel are more important to you.

An easy definition of the word "custom" is a procedure, a practice, a rule. Most of our customs are unwritten laws or regulations. "Courtesy" is best defined as your personal actions or attitude toward these practices. Most of these customs were established because of their steady use and practice in the everyday life of the military people.

As in your country, military rank possesses certain privileges dependent upon seniority. Subordinate officers and enlisted men are expected to exhibit an unfailing respect for the authority of rank. This does not mean that a senior officer may abuse those below him in rank. All of us must respect regulations, the rights of others, and act accordingly. Therefore, we respect rank by disciplined obedience combined with loyalty.

A very common custom is for a junior officer, or an enlisted man, to place himself to the left of his senior when in his company. You will notice also that during parades, ceremonies, and formal dinners the same rule applies. At receptions or social gatherings, officers remain at these functions until the commander has departed. The proper use of the word "sir" by a subordinate officer or enlisted man is imperative. It should precede a report, a question, and should follow the answer to a question. These are all examples that authority of senior rank is respected and recognized.

We also have customs regarding our officer and noncommissioned officer clubs. An officer is expected to join and support this activity. In this area, it is advisable to mention that all officers are expected to avoid excessive indebtedness by promptly paying financial obligations including club bills. A good credit standing is a must for any officer.

In our Armed Services, we have a code that is known as the "Officer's code." This code originated at the U. S. Military Academy long ago and is basically a philosophy of living. The code mentions that Duty is to be well performed, that we must have Honor in all things, that we must place country above ourselves. An officer's word is to be accepted, without question, as the truth. Lying and cheating are examples of violations of this code, and are punishable under the Uniform code of military Justice.

Airmen's Club

It is a custom as well as a requirement that officers present a good appearance at all times. The officer must set the examples which an enlisted man is to follow. Usually, the officers and enlisted men do not associate together. There are occasions such as participation on the unit athletic teams, or an official recreational function, when this is allowed. Normally, the officer and enlisted man go their separate ways.

The "salute" is an expression of courtesy and shows respect for rank and authority. The compliment of the salute is paid to all officers by enlisted men and by all junior officers to their seniors. This is probably done in your country also. We have noted, as you have, that this custom is observed by many Allied officers and enlisted men in respect to American officers. We would like to see this become an "international" custom, and the courtesy recognized by all students and American military personnel.

There are other occasions when the salute is rendered. Usually at 1700 hours each day, our flag is lowered. We call this "Retreat." At this time, if we are outdoors, and can see the flag or we can hear the music for "retreat," we face in the direction of the flag, salute, and hold that salute until the music stops. If we are in civilian clothes, we do not salute but we do stand at attention. If we are driving, it is necessary only to stop the car and sit still. There is also another occasion. During formal parades, we salute our flag as it passes by. We also salute when our national anthem is played.

The salute serves as an act of recognition between members of the military profession and as an indication of the possession of military courtesy and discipline by those who render it. Remember, we consider the salute as a sign of greeting, not of servility.

The commander desires to maintain a military atmosphere. It is an atmosphere we all understand. We know our objective and proceed to it in an organized and orderly manner. If you as a military individual understand this concept and act accordingly, we feel that you completely understand the most important of our military customs: respect for authority.

* Topic sentences

A headline in a newspaper, like a topic sentence, tells what the article (story) is about.

Example: CORPUS CHRISTI AREA HIT BY FLOOD WATERS

Moisture in the Gulf of mexico met a cool front over South Texas Tuesday, bringing flooding in the Corpus Christi area.

Heavy flooding along portions of Corpus Christi's beach combined with tides a foot above normal, bringing two inches of water into beachfront hotels.

High tides pushed water onto low-lying roadways, making some streets impassable.

Most paragraphs have something similar to headlines. . . .entence that tells what the paragraph is about is a topic sentence. It is usually, but not always, the first or last .entence in a paragraph.

Example:

Traveling by train is becoming more popular in the U. S. Because of the high cost of gasoline and the uncertainty of its supply, many people now prefer to travel by train. Some passengers even ship their cars by train, believing that it's cheaper to ship them than to drive them to their destination. When a person goes by train, he doesn't have to worry about the availability of gas, lodging, and food.

By reading the topic sentence, you know that the paragraph will be about train travel.

Re-read "A military Installation" and "Military Customs and Courtesies" and underline the topic sentence in each of the paragraphs.

* Role playing

Situation: A military (Air Force, Navy, Army) recruiter describes what a military installation offers and explains some of the military customs to a potential enlistee.

A recruiter from each branch could present the advantages of his particular service.

* Home Assignment

Choose the correct words from the list below to complete the following sentences.

slacks	slot	souvenir
slice	snack	spill
slippery	sour	squash

1. A wet fish is ______________ .

2. We had a midnight ______________ .

3. I bought a ________________ of New York when I visited there.

4. My mother used to grow ____________ in her garden.

5. Be careful. Don't ___________ Your coffee.

6. My new _______________ are made of wool.

7. We dropped our mail into the ____________ .

8. If you ___________ ink, it won't come off.

9. I don't like those grapes; they're ______________ .

10. Where can I buy a ________________ for my parents?

11. Could you cut me a ___________ of cheese?

12. ICE is ________________ .

13. Drop the coins in the ________________ .

14. We ate a small _________________ before lunch.

15. Lemons are ___________________________.

16. She made a sandwich using several slices of cheese between thin _______ of bread.

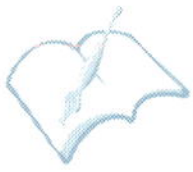

An American poem

Dreams

- Langston Huges(1902-1967) -

Hold fast to dreams
For if dreams die
Life is broken-winged bird
that cannot fly.

Hold fast to dreams
For when dreams go
Life is a barren field
Frozen with snow.

Lesson 9.

Julius Caesar

Wind Talkers I

Today's Clip

- You were right, Joe. I do look like a Nip.
- What the hell are you talking about?
- I'm putting on that Jap uniform, trying to get a hold of one of their radios, call off our guns.
- That's stupid.
- Shut up. Let him talk.
- He's not goin' out there.
- **You're not in charge here,** Enders.
- **I got orders,** Gunny. He's my responsibility!

Vocabulary Checkup

putting on	입다
get a hold of	연락하다
call off	물러나게 하다. 중지시키다
be in charge	~를 책임지다
orders	명령
responsibility	책임

Key Expressions

- **You're not in charge here,** Enders.
- **I got orders,** Gunny. He's my responsibility!

Let's Review!

A: Are you the person **in charge** of the equipment?

B: The person **in charge** happens to be away from his desk at the moment.

A: Who is in charge?

B: Sergeant Cage is in charge, sir.

Wind Talkers II

Today's Clip

- Enders.
- Yes, sir.
- The report I read claims **you went above and beyond.**
- Sir?
- **You've been recommended for commendation.** And **I concur.** But **I don't see any point in** you having to wait around for all those formalities to work themselves out. Not when I got one of these handy. Congratulations, Enders. You saved a lot of Marines.

Vocabulary Checkup

claim	의견
go above and beyond	일을 많이 하다.
recommend for	추천하다
commendation	상, 훈장
concur	일치하다
formality	정규절차

Key Expressions

- The report I read claims **you went above and beyond.**
- **You've been recommended for commendation.** And **I concur.**

Let's Review!

A: Do you know why the Captain Jones chose me for this job?

B: As far as I know, you were **recommended for** the job.

A: Do you know who recommended me?

B: I think Major Payne did.

READING

Cast of characters

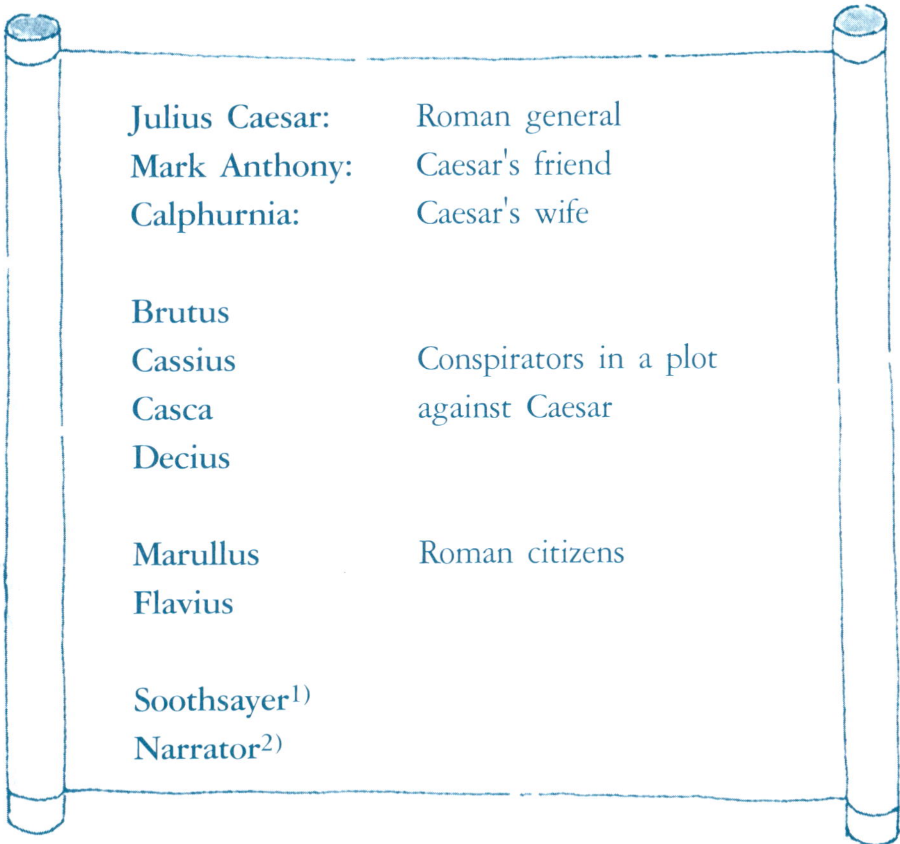

Julius Caesar:	Roman general
Mark Anthony:	Caesar's friend
Calphurnia:	Caesar's wife
Brutus	
Cassius	Conspirators in a plot
Casca	against Caesar
Decius	
Marullus	Roman citizens
Flavius	
Soothsayer[1]	
Narrator[2]	

ACT Ⅰ

[SCENE Ⅰ. – Rome. A Street.]

Narrator: Julius Caesar, a powerful Roman general, returns to Rome after winning a battle. Some of the people of Rome turn out[3] *to celebrate Caesar's victory, but others are angry and go about the streets*[4] *of Rome taking down flags and decorations*[5]*.*

1) soothsayer=fortuneteller
2) narrator: person who explains
3) turn out – 모이다 (=gather), 나오다 (= come out)
4) go about the streets – 거리를 돌아다니다. about=around

Marullus: Can we do this[6]? It is the feast of Lupercal[7].

Flavius: It doesn't matter. Let's have no statues hung with Caesar's trophies![8] We'll pull a few feathers from Caesar's wings! Then he can't fly so high.

Ⓠ What do Marullus and Flavius think of Caesar's victory?

[SCENE Ⅱ. - A Public Place.]

Enter Caesar, his wife Calphurnia, Anthony, Brutus and a great crowd who are on their way to see the holiday games[9].

Soothsayer: Caesar!

Caesar: I hear a voice from the crowd. Who is it? Speak. Caesar listens[10].

Soothsayer: Beware[11] the ides of March[12].

Brutus: A soothsayer bids you beware the ides[13] of March.

Caesar: Set him before me; let me see his face.

Julius Caesar

5) taking down flags and decorations 깃발과 장식물들을 끌어 내리면서
6) this = taking down flags and decorations
7) feast of Lupercal - 루퍼칼이라는 동굴 근처에서 해마다 2월 15일에 행해졌던 고대 로마의 축제
8) Let's have no statues hung with Caesar's trophies! - 어떤 조각상도 시저의 트로피로 장식되도록 내버려 두지 맙시다. hung with - …을 걸어서 장식하다.
9) the holiday games - 루퍼칼 축제의 경기들
10) Speak. Caesar listens. - 시저의 자기중심적이고 오만한 태도가 엿보인다.
11) beware: be careful
12) Beware the ides of March. - 3월 15일을 경계하라.
ides - (고대 로마력에서) 3월·5월·7월·10월의 15일 및 그 밖의 달의 13일.
13) ides[aidz]

Cassius: You, come from the throng; look upon Caesar.

Caesar: What say you to me now?[14] Speak once again.

Soothsayer: Beware the ides of March.

Caesar: He is a dreamer. Let's go.

Exit all but[15] Brutus and Cassius.

Ⓠ Was Caesar worried about what the soothsayer told him?

Narrator: Cassius is plotting to kill Caesar. He tries to convince Brutus to help[16] but Brutus is reluctant because he is a friend of Caesar's. However he does not want Caesar to be king.

Brutus: Listen! What is that shouting? I fear the people are choosing Caesar for their king!

Cassius: Do you fear it? Then I think you do not want him to be king.

Brutus: I don't, and yet[17] I love him well.

Cassius: I don't know how other people feel, but for myself I would rather not live than live in awe of someone no better than myself[18]. I was born as free as Caesar, and so were you[19]. We both have fed as well[20], and we can

---

14) What say you to me now? - 너는 지금 나에게 뭐라고 말하는가?
(= What do you say to me now?)
15) Exit all but ... - …을 제외하고 모두 퇴장 but=except
16) convince A to B - A를 B하도록 납득시키다.
17) and yet=but, nevertheless   I'm tired, (and) yet I can't sleep.
18) for myself I would rather not live than... 나로서는 …하느니 차라리 살지 않겠다.
in awe of - …를 두려워하며,…의 공포 속에서
no better than myself -나 자신보다 더 나을 바 없는
19) so were you = you were born as free as Caesar, too.

both endure the winter's cold as well as he.

Brutus: The crowd is shouting. I think they are giving Caesar some new honor.

Cassius: Men at some time are masters of their fates: The fault, dear Brutus, is not in our stars, but in ourselves, that we are underlings[21]. "Caesar"? Why has he grown so great? The wide walks of Rome now have room for only one man[22].

crown

Brutus: I have thought a lot about these things. And I will tell you something.

I would rather be a simple villager than live in Rome under the hard times that are coming[23].

Enter Casca.

Brutus: Tell us what happened today.

Casca: A crown was offered Caesar but he pushed it away; and then the people began to shout.

---

20) We both have fed as well (as he) feed=eat

21) = The fault that we are underlings is not in our stars, but in ourselves. 우리가 시저의 부하가 된 책임은 우리의 (운명을 결정하는) 별에 있는 것이 아니라 우리 자신에게 있다.
underling: person of lower rank or position
star (보통 -s) (사람의 운명을 결정하는 것으로 생각되는) 별

22) 로마의 넓은 보도는 이제 단 한 사람만을 위한 공간이 되었다. (여기서 one man은 시저를 가리킴.)
walks=roads

23) I would rather ... than ~ - ~하느니 차라리 …하겠다.
I would rather risk my life than be a coward.
under the hard times that are coming 앞으로 다가올 가혹한 시간들 속에서 (시저가 왕이 되면 전제 정치가 시작될 것임을 우려하고 있다.)

Cassius: They shouted thrice[24).

Brutus: Was the crown offered him thrice?

Casca: Yes, and thrice he refused it. But I thought he did not like to take his fingers off it[25).

Ⓠ Why do you think Caesar refused the crown thrice?

Act Ⅱ
[SCENE Ⅰ. - Rome.]

Enter Brutus in his orchard.

Brutus: I cannot, by the progress of the stars[26), give guess how near to day[27).

Enter Cassius and the other conspirators.

Cassius: Good morrow[28), Brutus. Do we trouble you?

Brutus: I have been up, awake all night.

Narrator: During the stormy night Cassius and the other conspirators convince Brutus to join them. They all decide to kill Caesar the next day.

---

24) thrice=three times
25) he did not like to take his fingers off it - 그는 그것(왕관)에서부터 손을 떼고 싶어 하지 않았다.
(즉, 왕관을 거절하고 싶어 하지 않았다.)
26) by the progress of the stars - 별들의 진행 상태로는
27) how near to day - 날이 새려면 얼마나 있어야 할지
28) Good morrow=Good morning.

[SCENE Ⅱ. – Caesar's House.]

Narrator: There are many signs and omens[29] warning of[30] danger. The next day Calphurnia asks Caesar to stay home.

Calphurnia: Horses did neigh, and dying men did groan, and ghosts did shriek[31] and squeal about the streets.

Caesar: These predictions are for the world in general[32].

Calphurnia: When beggars die, there are no comets; the heavens blaze for the death of princes[33].

Caesar: Cowards die many times before their deaths; the valiant[34] never taste of death but once[35]. It seems strange to me that men should fear. Death, a necessary end, will come when it will come[36].

Narrator: Decius comes to say the Senate is planning to offer Caesar a crown. If Caesar doesn't go, they may change their minds. Caesar wants to be king, and decides to go to the Senate.

---

29) omen=sign
30) warn of – …을 주의하다. …을 경고하다 I warned her of the danger.
31) shriek=scream
32) These predictions are for the world in general – 이러한 예언들은 대부분의 세상 사람들에게 해당되는 것이다.
People in general like new products.(=most)
In general people like new products.(=in most cases, usually)
33) no comets; (but) the heavens blaze...
34) valiant=brave ↔coward
35) the valiant never taste of death but once – 용감한 자들은 단지 한 번만 죽음을 맛본다. but=except
36) death=a necessary end – 필연적인 생명의 끝
Death will come when it will come. 죽음은 때가 되면 닥치게 마련이다. (즉, 누구나 언젠가는 죽는다.)

ACT Ⅲ
[SCENE Ⅰ. - Before the Capitol.]

On the way to the Senate, Caesar is walking with Brutus, Cassius, Casca, Decius and several others. Caesar sees the soothsayer.

As he was ambitious, I slew him.

Caesar: The ides of March are come.

Soothsayer: Yes, Caesar, but not gone.

Narrator: Just before the group reaches the Senate, the conspirators surround Caesar and stab him to death[37]*. As Caesar dies he says.*

Caesar: You too, Brutus? Then fall Caesar![38]

Cassius: Liberty! Freedom! Tyranny is dead!

37) stab him to death - 그를 찔러 죽이다.
38) You too, Brutus? Then fall Caesar! - 브루투스, 자네까지도? 그렇다면 시저는 죽는다! (절친한 친구로 믿고 있는 사람이 암살에 가담했기 때문에, 더욱 배신감을 느낀다.)

Casca: Cry out, "Liberty, freedom!"

Brutus: People and senators, do not be afraid. Ambition's debt is paid. No harm is intended to you. Let's all cry, "Peace, freedom, and liberty!"

[Scene Ⅱ. - Caesar's Funeral.]

Brutus: If anyone asks why Brutus rose against[39] Caesar, this is my answer: it is not that I loved Caesar less, but that I loved Rome more[40]. Would you rather have Caesar alive and be a slave, or have Caesar dead and be a free man[41]? As Caesar loved me, I weep for him; as he was valiant, I honor him; but as he was ambitious, I slew him.

Enter Anthony with Caesar's body. He speaks to the crowd.

Anthony: Friends, Romans, countrymen, lend me your ears[42]; I come to bury Caesar, not to Praise him. The evil that men do lives after them[43]; the good is often interred with their bones[44]; so let it be with Caesar[45].

39) rise against - (…에 반항하여) 들고 일어나다, 봉기하다.
40) It is not that I.... less, but that I ~ more. - 그것은 내가 시저를 덜 사랑했기 때문이 아니라 로마를 더 사랑했기 때문이다.
41) Would you rather have Caesar (be) alive... or ~ . - 여러분들은 시저를 살려 놓고 노예가 되기를 더 바랍니까, 아니면 시저를 죽여서 자유인이 되기를 더 바랍니까?
42) lend me your ears=listen to me
43) 악행은 죽은 후에도 남는다.
44) 흔히 선행은 뼈와 함께 묻힌다.[사람이 죽으면 그의 선행도 잊혀진다.]
45) so let it be with Caesar.=it will be so with Caesar. 시저의 경우도 그러할 것이다.

Brutus says he was ambitious, and Brutus is an honorable man[46]. Caesar brought much wealth to Rome[47]; did this seem ambitious? When the poor cried, Caesar wept; ambition should be made of sterner stuff[48]: yet Brutus says he was ambitious.

You all saw that on the Lupercal I thrice present-ed him a kingly crown, which he did thrice refuse[49]. Was this ambition? Yet Brutus says he was ambitious, and, sure, he is an honorable man[50]. I speak not to disprove what Brutus said, but I am here to say what I know.

You all loved him once. Why do you then not mourn him? *(Weeps.)* Bear with me[51]. My heart is in the coffin there with Caesar, and I must pause till it[52] comes back to me.

46) Brutus says he was... 브루투스는 시저가 야망이 있었다고 말한다. 그리고 브루투스는 거짓말 할 사람이 아니다. he=Caesar
47) Caesar brought much wealth to Rome. - 시저는 (많은 전쟁들을 승리로 이끌어) 로마에 많은 부를 가져왔다.
48) ambition should be made of sterner stuff (than weeping). - 야망은 우는 것보다 더 준엄한 것으로 만들어져야 한다. [그가 야망이 있었다면 울거나 하지는 않았을 것이다.]
49) which he did thrice refuse = but he did thrice refuse it.
50) Yet Brutus says he was ambitious... - 이 말을 반복하면서 결국 브루투스가 옳지 않았음을 시민들에게 암시하려 한다.
51) Bear with me - 내가 말하는 것을 참을성 있게 들으시오.
52) it=my heart

Anthony's funeral speech: "I come to bury Caesar, ..."

Narrator: *Them Anthony reads Caesar's will[53]. When the people learn that Caesar left most of his fortune to the citizens of Rome, they become very angry. They want to punish Cassius and the other conspirators, and rush off to burn down their house. Later a civil war breaks out between Brutus and Anthony. After months of fighting, Anthony is victorious, and Brutus is killed but Anthony mourns him and pays tribute[54].*

Anthony: This was the noblest Roman of them all[55]. All the conspirators save only he did that they did in envy of great Caesar[56]. He only, in a general honest thought

53) Caesar's will. - 시저의 유언장
54) pay tribute - 찬사를 보내다.
55) This was the noblest Roman of them all. - 이분(브루투스)은 모든 로마인들 중에서 가장 고귀한 분이셨습니다.
56) All the conspirators save only he - 그만을 제외한 모든 음모자들
did that they did in envy of great Caesar. 위대한 시저를 시기하여 그들이 했

and common good to all, made one of them[57]. But now let's call the field to rest[58], and let's away, to share the glories of this happy day[59].

by William Shakespeare

Think and Discuss

1. *Whom do you like best among the characters in this lesson?*

2. *Choose the sentences in this lesson that you like best. Why do you like them?*

던 것을 했다.(시저를 시기하여 죽였다.) that=what

57) He only, (in a general honest .... good to all,) made one of them. 브루투스만이 일반적인 정직한 생각과 모든 사람에게 공통의 선을 제공하려고 공모자들 중의 한 사람이 되었다.

58) let's call the field to rest - 전쟁을 끝내자. field=battlefield.

59) let's away, to share... 승리의 축배를 들자. (=let's go celebrate)

Reading Comprehension

A. Choose the unsuitable item from the list below to complete the main text.

a. full of awe
b. very stern
c. bloody and frightening
d. happy and humorous

B. Who spoke the following lines?

1. You too, Brutus? (   )
2. Beware the ides of March. (   )
3. I was born as free as Caesar, and so were you. (   )
4. It is not that I loved Caesar less, but that I loved Rome more. (   )
5. When beggars die, there are no comets; the heavens blaze for the death of princes. (   )
6. But now let's call the field to rest, and let's away, to share the glories of this happy day. (   )

C. True / False Questions.

1. ______ Julius Caesar was such a powerful Roman general that all the people of Rome wanted him to be king.
2. ______ The senators were so afraid of Caesar that they decided to slay him.
3. ______ The soothsayer's word proved to be right.
4. ______ Caesar left most of his fortune to the citizens of Rome.
5. _____ He didn't want Caesar to be king of Rome.

Let's Talk

I Wish I Were Anthony

Ingi and Subin are going to be in a play. They are waiting for friends at Subin's house.

Subin: How soon do you think they'll come?

Ingi: They should be here any minute[60] now.

Subin: I'm so excited! I've never been in a play before.

Ingi: We should practice our lines[61] while we're waiting.

Subin: Good idea. "Friends, Romans, countrymen, lend me your ears."

Ingi: Those aren't your lines! You're playing Cassius.

Subin: But I wish I were Anthony. I'd like to be the hero.

Ingi: I'm not sure there is a hero. But Anthony certainly is on stage more than anyone else.

Subin: That's exactly what I had in mind.

Ingi: Ah! Subin is an ambitious man!

Subin: Cut that out[62]! I really just like Anthony's lines.

60) should be here any minute - 곧 올 것이다.
61) practice one's lines - 대사를 연습하다.
62) Cut that out! - 그 말은 빼라! 그만해!

Lesson 10.

The Nautilus-The First Atomic Submarine

Watching Movie

COURAGE UNDER FIRE I

Today's Clip

- It was, uh... unusual, wasn't it, for Karen to want to be a helicopter pilot?
- **I guess that's my fault.** I took her to the fair when she was 11.
- 12... 7th grade.
- Anyway, there was a helicopter there. Small one with the bubble up front. The fella the pilot was selling rides. I bought Karen a ticket. It was a short ride. But when she came down... Well, **I don't think she ever** came down, actually.

Vocabulary Checkup

unusual	보기 드문
fair	박람회
bubble	거품
fella	(fellow) 사람, 녀석
ride	ⓝ 탈것

- **I guess that's my fault.** I took her to the fair when she was 11.
- Well, **I don't think she ever** came down, actually.

Let's Review!

A: Where is James again? **I don't think he ever** does his job. He is such a pain.

B: **I guess that's your fault.** You always let him go home whenever he wants.

A: I just wanted to be nice.

B: Yeah, but now he thinks he can go home when he wants, even if he has work to do.

COURAGE UNDER FIRE II

Today's Clip

And so, ladies and gentlemen, **what I have to say to you** in closing is to go forth, to command wisely, to advise prudently, to protect with vigilance, and to judge with care. And **above all**, to defend with all your heart, all your soul, and all your courage the lives of the men and women under your command, and the United States of America.

Vocabulary Checkup

forth	앞으로
prudently	신중하게
vigilance	경계
defend	방어하다

Key Expressions

- And so, ladies and gentlemen, **what I have to say to you** in closing is to go forth, to command wisely, to advise prudently, to protect with vigilance, and to judge with care.
- And **above all**, to defend with all your heart, all your soul, and all your courage the lives of the men and women under your command, and the United States of America.

Let's Review!

A: Is there any way I can ever master Chinese? It's so difficult.

B: **What I have to say to you is** that learning a language is a life-long process.

A: I know, but how do you speak Chinese so fluently? What's your secret?

B: I don't have one. I just studied hard.

COURAGE UNDER FIRE III

Today's Clip

- Come on, Captain. I don't want to wait here to die.
- Ilario?
- I don't know.
- Come on, it's a majority.
- **That would be great if** this was a democracy, but it isn't. We stay with Rady.

- **I wouldn't** risk your life. I won't risk his.
- Okay. So, maybe if we surrender, the Iraqis will doctor him up.
- No surrender!

Vocabulary Checkup

majority	과반수
democracy	민주주의
risk	위태롭게 하다
surrender	항복하다
doctor ___ up	~를 치료하다

Key Expressions

- **That would be great if** this was a democracy, but it isn't. We stay with Rady.
- **I wouldn't** risk your life. I won't risk his.

Let's Review!

A: **It would be great if** the test is next week.

B: But it's tomorrow! Aren't you ready?

A: Whew~ Not at all. Would you study with me tonight to cram for the test?

B: Sorry, but I would rather have a good night's sleep before the test.

READING

On January 17, 1955, the first nuclear-powered[1] submarine[2] went to sea. The Nautilus, an atomic[3] submarine of the United States Navy, was the first true submarine. Other submarines, prior[4] to that date, were actually surface ships that were able to go under water for only short periods of time.

In the spring of 1953, work on the Nautilus was begun. One year, seven months, and seven days later, at a cost of $29,000,000, the first atomic-powered submarine was completed[5]. This was only the submarine, not the engine. The nuclear-powered engine was installed[6] a year later.

1) Nuclear powered: operated by the use of atomic energy
2) Submarine: a ship that can operate below the surface of the water
3) Atomic: using power or energy from unclear chain reaction
4) Prior: before, earlier in time
5) Complete: to finish or be through with something
6) Install: to fix in position for use

The men of the navy and the atomic scientists were not actually sure an atomic engine could safely be used in a submarine, but work began on the submarine while experiments were still going on with nuclear power. The submarine was to be constructed[7] in one place, and the engine constructed in a research laboratory[8] at another location.

The initial[9] step in the building of a nuclear submarine was to design[10] and plan what it would look like. An exact solid[11] wooden model was made. It was made to scale, with one inch equal to one foot. Next, after this model[12] was carefully studied, a full size wooden model was constructed. This model showed where all of the pipes, valves, instruments, etc., were to be located. From this full size wooden model, patterns[13] were then made. These patterns were for the steel plates to be used in making the hull[14] of the ship. This first atomic submarine was to contain more than 7,000,000 steel plates. The steel used for the hull was over 2 inches thick. It was so thick that the usual tools could not penetrate[15] it. So, the steel had to be cut with a special burning torch. The steel sheets were curved to form the shape of the hull and were welded[16] together.

---

7) Construct: to build or form something from parts
8) Laboratory: a place or room for research or experiments
9) Initial: the first of something
10) Design: to sketch or make preliminary plans
11) Solid: something having length, breadth, and thickness; not hollow
12) Model: a copy or imitation of an object usually made to scale
13) Pattern: a guide or model used to make something
14) Hull: the frame or main body of a ship
15) Penetrate: to pierce or pass through with any sharp object
16) Weld: to join metal together with heat

The propellers[17] were made of solid brass and weighed 5 tons. They had to be perfectly balanced. If they had not been, the vibration[18] from the propellers would have shaken the entire ship. The propellers were so accurately balanced that a man could rotate the blades with his finger.

Submarines have to be constructed with two hulls. The crew lives and operates the submarine from the inner hull aboard[19] the ship. The crew's quarters[20] contain the living and sleeping areas. The engine, torpedoes,[21] etc., are also in this inner hull. This hull must be completely watertight[22]. If the outer hull is not watertight, it will leak. The outer hull contains the ballast tanks. the ballast tanks[23] make it possible for a submarine to dive under the water or to come to the surface. When a submarine dives, valves are opened and the ballast tanks are filled with water. Water in the tanks is blown out with compressed air[24] and submarine floats to the surface.

17) Propeller: a large twisted blade that rotates and moves a vessel through water
18) Vibrate: to swing or shake back and forth
19) on or in a ship or airplane
20) a military term for a living and sleeping area
21) metal cartridges launched from submarines which explode on contact with objects
22) constructed so tight and snug no water can get in
23) a metal tank carried on ships to give stability

Work continued 24 hours a day, 7days a week on the construction of the Nautilus. All of the metal used throughout the submarine was thoroughly examined and tested many times for any defects.[25] The metal was tested under pressure, heat, and cold. All of the parts and joints were X-rayed in order to detect any possible leaks or flaws in the material or the construction. any defective part was rejected. The pipes used on the ship were made of copper, brass, steel, and aluminum. There was 40 miles of electrical wiring used on this first atomic submarine.

On January 21, 1954, the Nautilus was launched and the navy took cover[26] command. The atomic engines had not yet been installed. Work was finally completed on January 17, 1955, and the Nautilus departed on its first trial run[27] under full nuclear power. Numerous records were broken on the first cruise. It was able to travel 1,300 miles submerged,[28] taking 84 hours. It was able to maintain an average speed of 16 knots (18 1/2 miles per hour).

The main purpose of a submarine is to fire torpedoes at enemy ships.

24) air reduced in volume by pressure and held in a container
25) deficiency, imperfection, flaw
26) assumed command or power
27) experimental trip or operation conducted under controlled conditions
28) to place under or cover with water

And the Nautilus carried 24 torpedoes for 6 tubes. Each torpedo was 21 feet long and weighed 3,000 pounds. The ship had a crew of 100 men to operate all of the equipment and to run the submarine. These men had to be trained[29] at a special school.

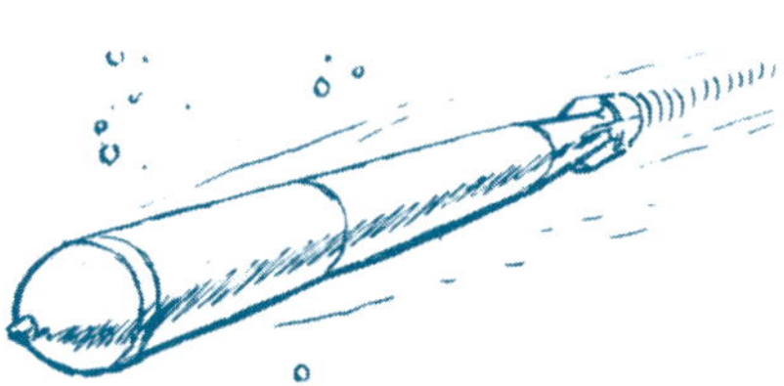

Class questions and discussion topics

1. Why do you think men were so eager to begin construction of an atomic submarine in 1953, when the final designs were not even completed?

2. What do you believe are some of the advantages of a ship that can stay submerged for long periods of time?

3. Why does a propeller on a ship or aircraft need to be accurately balanced?

4. Why do you think it was necessary to make an exact scale model of the Nautilus before actual construction was begun?

29) to guide, teach, or instruct

5. How is it possible for one man to rotate a 5-ton propeller?

6. Why do you think the different metals used throughout a submarine must first be exposed to freezing conditions and then to intense heat?

7. Why do you believe it is so critical to locate any type of leak in the hull of a submarine?

8. Describe the process a submarine uses to come to the surface.

9. Describe what you might see if you walked through the crew's quarters on a submarine.

10. Why was the Nautilus able to stay submerged for such long periods of time?

EXERCISE 1

Time and Date Drill "on, in, at"

a. The preposition *on* is used in the area of time and dates to give specific or exact dates, such as:

on Wednesday. *on* January 14th.

b. *In* is used to indicate a duration or period of time, but no exact date or time is given, such as:

in March. *in* the first week....

c. The preposition *at* is used in a time situation to indicate an exact hour or minute of time, such as :

at 2:30. *at* noon.

*** Select the correct word to complete the sentences below.**

1. They began construction _____ the 15th of June.

2. You can use the room ____ Wednesday, but not ____ Tuesday.

3. They began construction early _____ June.

4. Class will not begin _____ 1:30 today, but _____ noon.

5. I have to be there _____ midnight.

6. We'll leave _____ Tuesday.

7. The submarine left _____ January 17th _____ 5 P. M.

8. I think we can finish this unit _____ two hours.

9. We can see the result of our work _____ a day or two.

10. There won't be enough time _____ Monday for a test.

11. Since there's no school today, I'll meet you ____ lunch time near the library.

12. I'll be finished with my class _____ Friday if I pass my test.

13. He should be here _____ a minute.

14. _____ two weeks they will be completely finished.

15. They told me to be in the lab _____ 4:45 to start the test.

EXERCISE 2

Select the term with the same meaning from the list below to replace the italicized word(s) in each of the following sentences.

prior to	vibration	design
accurately	patterns	construct
model	install	initial
penetrate	completed	solid

1. You can use a sharp knife to *pierce* the material.

2. The engineers were going to *build* a new type of submarine.

3. The *first* step was to make a scale model of the ship.

4. He wanted to *plan* a new type of machine that required no maintenance.

5. They had to *precisely* measure every steel plate that was to be used.

6. The project was *finished* on the 18th day of last month.

7. He tried to *put in* the new motor but he didn't have the right tools.

8. He carved an exact *copy* out of wood.

9. She was terrified as she felt the *shaking* from the earthquake.

10. He wanted all of us in the classroom 10 minutes *before* the test.

11. The doors are so heavy because they are *not hollow*. They're made of oak.

12. She made some *models* out of paper before she started to cut the dress material.

HOMEWORK ASSIGNMENT

Ⅰ. COMPLETION EXERCISE

Write the word in the blank space that correctly completes the sentence.

designing	provisions	construct	accurately
initial	on	aboard	defects
departed	around	detect	penetrate
float	quarters	examine	trapped
laboratory	leaking	completed	

1. If you come ________ the ship, I'll show you around.

2. There was no way to __________ count all of the work that had been done.

3. The crew will begin work ________ the first Monday of next month.

4. He thinks he will have all of his work ________ by midnight.

5. I can't accept this material. There are a lot of ________ in it.

6. He's using wood and plastic to ___________ his model plane.

7. The train arrived at 6:16 and ____________ at 6:45.

8. He's ____________ a new office building that will be made of glass and steel.

9. I think there is something wrong. This machine will be able to ____________ the problem.

10. You will have to ____________ every part very carefully.

11. I hope this rubber raft will ____________ .

12. There are many things to remember. But the ________________ thing is to turn the engine off.

Humor English

Something with diamonds in it

Sam asked his wife what she would like for her birthday, and she said to him, "I'd like something with diamonds in it."

So he bought her pack of cards!

다이아몬드가 들어 있는 것

샘이 부인에게 생일 선물로 무엇을 원하느냐고 묻자 그녀가 말했다.

"다이아몬드가 들어 있는 것을 원해요."

그래서 그는 그녀에게 카드 한 벌을 사주었다!

Lesson 11.

Training in Space

Watching Movie

THE THIN RED LINE I

Today's Clip

- So **I'm asking for volunteers.** Go back up there and knock it out.
- Sir? I'll be glad to go back up and lead the way back for a party.
- All right. That one.
- I'll go, sir.
- You. All right.
- I'd like to lead the party, Colonel.
- All right, John. **I was hoping you'd say that.** Go ahead and you'll be in command.
- I'll go, sir.
- Alright. You.
- Sir?

- You.
- All right, men, that's enough. Seven is plenty. Gather up.

Vocabulary Checkup

back up	후위를 맡다
knock out	패퇴시키다
party	부대
plenty	충분한
gather up	모으다

Key Expressions

- So **I'm asking for volunteers.** Go back up there and knock it out.
- All right, John. **I was hoping you'd say that.** Go ahead and you'll be in command.

Let's Review!

A: What are you doing here?

B: I'm posting an ad.

A: **I'm asking for volunteers** to help the disabled people.

B: Can I join?

A: **I was hoping you'd say that**. Thanks.

THE THIN RED LINE II

Today's Clip

- What're you doing down here?
- Just···you know, **I'm heading back to** the company.
- Yeah. Well, you give all them boys my best.
- **You can come with me if you want.** I'll help you along.
- No, it's nice and quiet. Peaceful up here.

Vocabulary Checkup

head	(~를 향해) 가다
company	중대
along	(~와) 함께
quiet	조용한
peaceful	평화로운

Key Expressions

- Just···you know, **I'm heading back to** the company.
- **You can come with me if you want.** I'll help you along.

A: Hey, Jane! Where are you going?

B: **I'm heading back to** the office. I forgot to bring something I need.

A: Really? I'm driving that way too. **You can come with me if you want.**

B: Oh, sure. Thanks.

READING

A space station or space laboratory is an artificial[1] satellite[2] that orbits through space. Men are able to live and to work in a space station. It is a permanent[3] structure. This type of laboratory can be constructed to orbit without any possibility of reentry into the Earth's atmosphere. It is up in space for good[4]! In fact, if a space laboratory were to enter the pull[5] of Earth's gravity, it would be completely destroyed[6]. Astronauts and scientists are to live and work in these space labs for extended periods of time. The means of transportation to a space laboratory will have to be some type of spacecraft that can be lifted from earth into space and maneuvered[7] to the station. These space stations can be of great importance to astronautical engineers. They will be utilized[8] as "on the spot" laboratories. Experiments will be conducted, testing life support equipment for future space travel. The space labs will also provide places to test lunar[9] interplanetary space vehicles[10] and

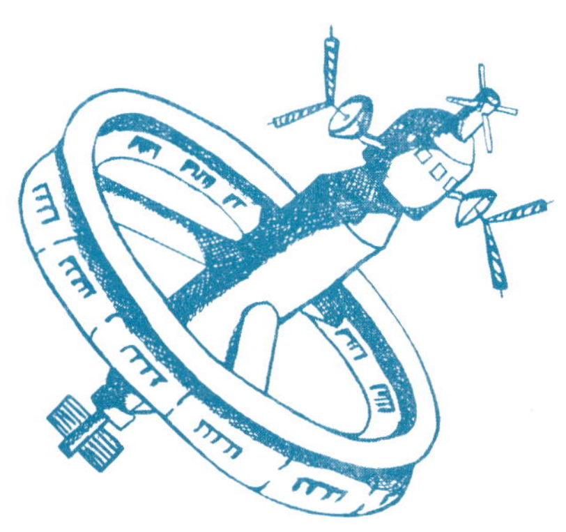

1) Artificial: simulated, and imitation of something natural
2) Satellite: an object orbiting the earth, moon, or other planet
3) Permanent: lasting, not temporary
4) For good: permanently, always
5) Pull: drag or draw
6) Destroy: to demolish, break up, or ruin
7) Maneuver: to perform movements according to a procedure
8) Utilize: to put into use, or use
9) Lunar: of the moon, or related to the moon
10) Vehicle: any device used for carrying persons or objects

other types of equipment to be used in future lunar landings.

One area of major[11] importance is that space laboratories will serve as simulated[12] training stations for astronauts. This will enable them to prepare for long space missions. A space laboratory can solve another big problem. In its orbital cycles, it will be in a state of weightless-ness. Weightlessness can be created on earth, but not for adequate[13] periods of time. Space stations can be utilized for biological experiments in which scientists conduct special types of experiments at zero or low gravity levels. The psychological effects of space travel can be brought into direct application by having the astronauts live for a specified period of time in the close confinement[14] of the space lab. The mental attitude of the men, their efficiency[15] capabilities[16], endurance[17], and degree of teamwork[18], can all be determined by the trial living period within the space station. Rather than[19] try to simulate a space trip on earth, a compulsory[20] training period in the lab of several days to several weeks could prepare an astronaut for a successful space trip of many months or even years.

11) Major: greater in rank or importance, utmost
12) Simulate: to pretend, imitate
13) Adequate: enough, sufficient
14) Confine: to keep within limits, enclose
15) Efficiency: the ability to produce or perform with a minimum of effort
16) Capability: ability not yet developed or utilized
17) Endurance: the ability to last, continue, or remain
18) Teamwork: joint action by a group working together
19) Rather than: instead of
20) Compulsory: required

Human survival in space must depend on well-developed and reliable life-support systems. Everything that makes up man's daily life cycle must require some adaptation[21] to space existence. Eating, bathing, sleeping—all of these must be reevaluated and adjusted[22] to space travel. Small space stations can be assembled[23] and fully equipped on the ground They can then be launched into orbit as permanent structures, never to return to earth. The space station must rotate as it orbits through space in order to provide a certain amount of artificial gravity. It is usually constructed in two sections. A support system is located in the rear and the laboratory or working area in the front part. The laboratory will usually be constructed slightly off center to minimize the radius of rotation. Larger space stations are usually round wheel-shaped structures. This shape avoids the difference in the distance between the life-support system and the laboratory area.

The astronauts that will man and train in a space laboratory will be transported there in space vehicles similar to those used in the Apollo and Gemini missions. These ships will contain rockets powerful enough to lift and carry them into orbit. The astronauts that traveled in the Apollo and Gemini capsules were well trained in two major procedures related to the use and manning of space laboratories: docking and rendezvousing maneuvers. Docking is joining[24] two objects in contact with each other and then attaching or coupling[25] them together. To

21) adjustment to something
22) to regulate
23) to put together or bring together
24) bringing together, attaching
25) join together, link, or connect

rendezvous is to meet another body in orbit, such as another space station or laboratory.

Class Questions and Discussion Topics

1. Which do you believe would be of greater value to future astronauts, a long technical training program on earth, or a short trial period aboard an orbiting space lab, and why?

2. Since orbiting space stations have become such great assets in the scientific field, what nontechnical areas do you think could utilize orbiting satellites?

3. Why do you think experiments in isolation and close confinement cannot be as successfully carried oe ain ground simulated situations ain hey canain space laboratory experiments?

4. What psychological problems, do you believe, can be detected in a simulated space flight experiment?

5. Who would be better suited for and able to maneuver in a weightless state, a very large man or a small man?

6. What specific type of equipment, do you think, must be an essential part of the life-support system?

7. Why are space stations continually orbiting the earth instead of being permanently located in space?

8. Which would be more essential to a space flight crew whose mission included a docking and/or rendezvous, a mathematical or engineering background?

9. What civilian use, if any, can be made of the electronic and technical equipment used for docking procedures in space?

10. Do you consider space travel very safe or still relatively dangerous?

EXERCISE 1

Select the term with the same meaning from the list below to replace the italicized word in each of the following sentences.

lunar	for good	maneuvered
utilize	permanent	pull
destroyed	artificial	

1. The records of the space flight would be a *lasting* record of the mission.

2. The scientists are planning two unmanned *moon* flights next month.

3. The smaller space capsule allows the astronauts to *use* visual docking procedures.

4. Scientists have been able to create a *simulated* laboratory on the ground.

5. A number of space laboratories would be *demolished* if hey entered the earth's atmosphere.

6. But a few newer ones can be *manipulated* to reenter this atmosphere.

7. The original idea was to send these space stations into space *permanently*.

8. Scientists finally found something that could withstand the *drag* of gravity as the capsule orbited.

EXERCISE 2

Select the term with the same meaning from the list below to replace the italicized word(s) in each of the following sentences.

several	compulsory	adequate
limited	rather than	simulate
major	capability	

1. He wants coffee *instead of* tea for his breakfast.

2. They tried *a few* experiments alone, but found that teamwork produced the best results.

3. He wondered if he had the *ability* for this type of job.

4. They were *confined* to the laboratory area during working hours.

5. This probed to be a very *important* project in the space program.

6. There is not *enough* space in the laboratory for three astronauts to move around in, with ease.

7. They were able to *pretend* all the actions of a real lunar landing.

8. Survival training is *required* for all astronauts.

HOMEWORK ASSIGNMENT

Ⅰ. SENTENCE FORMATION EXERCISE

Combine the two statements into a single sentence as shown in the example.

Example: She enrolled in a course. She wanted to improve her typing.
She enrolled in a course to improve her typing.

1. I'll shut the door. I want to keep out the noise.

___

2. The men walked fast. They wanted to get there on time.

___

3. He want back later. He wanted to see the statue again.

___

4. He questioned us. He wanted to find out the answer.

___

5. They went to the store. They wanted to buy some cigarettes.

___

6. Her mother gave her a list. She wanted to remind her.

____

7. We stayed at home. We wanted to finish the report.

____

8. He's going to open the window. He wants some fresh air.

____

9. I went to bed early. I wanted to get more rest.

____

10. He borrowed some money. He wanted to pay for the gas.

____

11. He turned on the TV. He wanted to see a program.

____

12. I've taken an examination. I want to qualify for the job.

____

Humor English

Panda

A panda walks into a bar, sits down, and orders a pint of draft beer. He drinks it, pulls out a gun, and shoots the waiter seriously hurt.

As the panda stands up to go, the bartender shouts,"Hey! Where are you going? You just shot my waiter badly injured, and you didn't pay for your drink."

The panda yells back at the bartender, "Hey man, I 'm a panda. Look it up!"

The bartender consults his dictionary to "panda" and reads.

"A tree-dwelling marsupial of Asian origin, characterized by distinct black and white coloring. Eats shoots and leaves."

단어

draft beer: 생맥주. look up: (사전을)찾아보다(consult).

marsupial: 유대류(有袋類). shoot: 쏘다, (식물의) 싹.

leaves: 떠나다, 잎(leaf)의 복수.

Lesson 12.

If I were seventeen again

Watching Movie

BORN ON THE FOURTH OF JULY I

Today's Clip

I had a dream, Ronnie, the other night, and you were speaking to a large crowd just like him. **Just like him**. And you were saying great things.

My fellow Americans, ask not what your country can do for you, ask what you can do for your country.

Vocabulary Checkup

speak 연설하다
crowd 군중, 대중
fellow 동지

Key Expressions

- **I had a dream**, Ronnie, the other night, and you were speaking to a large crowd just like him.
- **Just like him**. And you were saying great things.

Let's Review!

A: Hello? Mom, is Dad there? Is he OK?

B: Why wouldn't he be OK?

A: Well, I had a dream about him last night. He was in the hospital.

B: He's fine! You know what? You're just like him. He gets worried from his dreams too.

BORN ON THE FOURTH OF JULY Ⅱ

Today's Clip

First off, young men, **let's get one thing straight**. Not everybody becomes United States Marines. We want the best, and we'll accept **nothing but the best**, because there is nothing prouder, nothing finer, nothing standing as straight as a United States Marine. Now, they've got the Army. They got the Navy. They got the Air Farce. And if you want to join them, be my guest.

Vocabulary Checkup

straight	바로, 올바로
prouder	더 자랑스러운
finer	더 멋진
be my guest	환영하다, 마음대로 해라.

Key Expression

- First off, young men, **let's get one thing straight**.
- We want the best, and we'll accept **nothing but the best**,

Let's Review!

A: The soccer game is tomorrow. Are you guys ready?

B: Yeah. We did a lot of training. I think we are ready.

A: Just do your best, **nothing but the best**.

B: We will! Trust us.

Mini Dialog A

What do the students dream about?

Alice dreams about living in the country. Fred dreams about owning a dog. Nancy dreams about leading her class[1]. Sam dreams about being a great athlete.

What would you do if[2] you were rich?

If I were rich, I would travel around the world.

How about your friends?

If Bill lived near a lake, he would go fishing.

If Susan had a job, she would save all her money.

What would you like?

I would like space to hunt on[3].

I would need a farmhouse to live in.

Tell me, what would you like to have?

I would like a red coat, but a blue one would do.

Would you like to have a car?

Why not? But a bike would do[4].

1) lead one's class - 반장을 하다.
2) What would you do if...? - 만을 … 라면 어떻게 하시겠습니까?
3) space to hunt on - 사냥을 할 수 있는 터.
4) a bike would do. - 자전거만 있어도 괜찮습니다. 여기서 do는 [족하다; 도움이 되다]의 뜻.

Mini Dialog B

Which is more important, grades or character?

The latter[5]. An honest character is very important.

Which is easier, school work or farm work?

The former. Farm work takes great physical strength.

What should I keep in mind?[6]

No matter how rich you are, you can't buy everything.

No matter what happens, let's stay friends.[7]

No matter where you go, stay in touch with[8] me.

It doesn't matter whether you get A's or B's on the test.

It is better to have a good character than to[9] have good grades.

You must consider it a duty to serve in the army.

Try to have a reputation for honesty.

What are you reading, Sally?

A book about nature. I'm trying to get acquainted with trees and animals.

5) The latter – 후자. 반대는 the former (전자).
6) What should I keep in mind? – 내가 명심해야 할 것은 무엇입니까?
7) Let's stay friends. – 우리 계속 친구로 지내자.
8) stay in touch with – ~와 연락을 하고 지내다.
9) It's better to ~ than to... – … 하는 것보다 ~하는 것이 낫다.

READING

If I were seventeen again, I would want to live on a Kentucky hill farm. I would want to grow up where there are trees, meadows and streams. If I couldn't live on a large farm, a few acres would do. But I would want space to hunt on, with a stream or lake nearby where I could fish.

In the spring of the year, I'd want to take long walks into the woods. I'd get acquainted with[10] all kinds of birds. I'd learn how they build their nests, what color and size of eggs they lay[11], and how they feed their young[12]. I'd want to know all about the animals. I would find out what they eat, where they live, which are friendly and which are enemies. I would also learn the names of the trees, plants and flowers that grow in the woods.

10) get acquainted with - ~과 친해지다.
11) what color and size of eggs they lay - 새들이 어떤 색깔과 크기의 알을 낳는지.
12) feed one's young - 새끼에게 먹이를 주다.

I would want to won at least one dog and go hunting with him. A boy of seventeen who[13] has not stood on a high hill under the stars or a bright moon and listened to the music of barking dogs[14] has missed something really great. Such an experience gives him something he cannot get anywhere else[15].

In summer, I wouldn't miss working on a farm. I wouldn't miss plowing land, planting seeds, and growing young green corn. I would mow the meadows and carry the hay to the barn on a hay wagon. If anyone hasn't smelled newly-mown hay[16], he has missed the finest smell[17] a youth ever breathed. I would want to work in shorts, without a shirt and with bare feet[18]. The feel of warm soil under one's feet is a good thing in one's growth[19].

If I were seventeen again, I'd try to make my body strong[20]. I wouldn't smoke, or drink anything bad[21]. A young man takes pride in strength and he can build strength by proper work and good exercise.

And there is something I would definitely do: I'd go to high

13) A boy of seventeen who – …한 17세의 소년. 동사는 7행의 has missed.
14) the music of barking dogs – 개 짖는 소리, 그 소리를 music (음악)이라는 단어로 미화했다.
15) something he cannot get anywhere else – 다른 곳에서는 전혀 얻을 수 없는 어떤 것.
16) newly-mown hay – 갓 베어낸 꼴 / 목초.
17) he has missed the finest smell... – 그는 젊은이가 맡아 본 냄새 중 가장 순수한 냄새를 놓친 것이다.
18) with bare feet – 맨발로.
19) in one's growth – 성장기에.
20) make one's body strong – 몸에 튼튼하게 하다.
21) ... drink anything bad – 뭐든 좋지 않은 것은 마시지 않겠다. drink 앞의 I wouldn't에 연결된다.

school. The boy who hasn't finished high school has missed something. It doesn't matter whether he leads his class or not, whether he's the best athlete or the most popular boy.

I never had these honors, but studying different subjects under different teachers, and knowing the boys and girls in my class are things I wouldn't miss[22]. I would consider it enjoyment to[23] attend high school. If I couldn't be the best athlete, I would still be one[24]. I wouldn't miss athletics. This builds men physically and teaches sportsmanship and how to give and take[25].

If I were seventeen again, I would earn my own money. If I had a job, I would do it so well that my employer[26] would want me to work for him again. Do work well at seventeen and you'll be doing it well from then on[27].

22) things I wouldn't miss – 놓치고 싶지 않은 것들.
23) consider it enjoyment to... – …하는 것을 즐거움으로 생각하다.
24) I would still be one – 그래도 운동선수가 되겠다.
25) give and take – 주고받다; 양보하다.
26) I would do it so well that my employer... – 일을 잘 해서 고용주가 …하게끔 하겠다.
27) from then on – 그 후로 계속해서.

I would want a reputation for honesty. If I borrowed money from someone, I would pay it back. If I were to choose[28] between a strong body and a strong, honest character, I'd take the latter. It is better to have C grades in school[29] and an A character than to have A grades and a C character.

If I were seventeen again, I would stand up for[30] my beliefs. I wouldn't be a follower of something that I didn't believe in or something that I knew was wrong. No matter how popular the idea might be, I would hold out[31]. I would be myself[32], and would be guided by what I thought was right[33]. No one, no matter how much more powerful he was than I, would sway me.

Seventeen may be the shortest year in your life. It was for me. It was a wonderful year and, like a Kentucky April, it came too slowly and went too quickly. It was a great year for physical and mental growth, a year of beauty and spirit.

28) If I were to choose... – 내가 만일 …을 선택해야 한다면.
29) have C grades in school – 학점을 C를 받다.
30) stand up for ~ – 옹호하다 ; 지지하다.(=support)
31) hold out – 버티다; 견디다
32) I would be myself – 나는 주체 의식을 갖겠다.
33) what I thought was right – 내가 옳다고 생각한 것.

Reading Comprehension

A. True / False Questions.

1. The author would be satisfied with a few acres. T/F
2. He would try to learn about the habits of animals. T/F
3. He would like to own a cat as a pet. T/F
4. He would not wat to work on the farm. T/F
5. It is necessary to be an outstanding athlete. T/F
6. He would want a reputation for being honest. T/F
7. Good grades are better than a good character. T/F
8. He would be guided by what he thought was right. T/F

B. Select the unsuitable item from the list below to complete the main text.

1. The author would want to live on a farm.
   (a) He would like to go hunting and fishing.
   (b) He is interested in learning about plants and animals.
   (c) He would like to have at least one dog.
   (d) He would not enjoy plowing land and planting seeds.

2. He would try to develop his body.
   (a) He would not stat to smoke cigarettes.
   (b) He would avoid drinking anything bad.
   (c) He would never eat fast food.
   (d) He would build strength through work and exercise.

3. He thinks a high school education is very important.

   (a) Everyone should try to lead his or her class.

   (b) It is good to study different subjects with different teachers.

   (c) Students should try to get to know their classmates.

   (d) School athletics are an important part of growing up.

An English poem

Down by the Salley Gardens

- W. B. Yeats(1865-1939) -

Down by the salley gardens my love and I did meet;
She passed by the salley gardens with little snow-white feet.
She bid me take love easy, as the leaves grow on the tree;
But I, being young and foolish, with her would not agree.

In a field by the river my love and I did stand,
And on my leaning shoulder she laid her snow-white hand,
She bid me take life easy, as the grass grows on the weirs;
But I was young and foolish, and now am full of tears.

* salley n. 갯버들 sallow의 속어.

Lesson 13.

Military Briefing: making power point

Watching Movie

G. I. JANE I

Today's Clip

- Lieutenant O'Neil reporting for training, sir.
- Yes, of course.
- Please, Lieutenant, **have a seat**.
- Thank you, sir.
- **Would you care for** a beverage? Tea? Soda?
- I'm fine, sir.
- You sure?

Vocabulary Checkup

Lieutenant	중위
report	보고하다
care for	~하고 싶어하다

beverage 음료수
sure 확신하다

Key Expressions

- Please, Lieutenant, **have a seat**.
- Thank you, sir.
- **Would you care for** a beverage? Tea? Soda?

Let's Review!

A: Ah, there you are! Come in, come in. Have a seat.
B: Thank you.
A: **Would you care for** refreshments? I have water and coke.
B: Coke would be good. Thanks.

G. I. JANE II

Today's Clip

In spite of your expertise, many of you will not successfully negotiate this program. Those that do will experience an operational tempo that far exceeds that of any unit within the U. S. arsenal. That is all that is to be said... about the special nature of the C.R.T. training program.

Vocabulary Checkup

negotiate 수행하다
operational tempo 작전 템포

exceed 넘다 능가하다
arsenal 무기고
C.R.T. training program CRT(Combined Reconnaissance Team) 합동수색팀 훈련 프로그램.

Key Expressions

- **In spite of** your expertise, many of you will not successfully negotiate this program.
- Those that do will experience an operational tempo that far exceeds that of any unit within the U. S. arsenal.

Let's Review!

A: Was Private Kim wearing green socks with his PT gear this morning?

B: Yes, Sergeant, he was.

A: He's asking for it. He told me yesterday he thought the new regulation prohibiting green socks with PT gear was stupid. He was going to wear green socks on purpose, **in spite of** the new regulation.

G. I. JANE III

Today's Clip

- Check out who's back, man.
- Mr. Wickwire !
- Yes, Master Chief ?

- Since the whole world seems to be awake at 0500, **why don't you** get your class out here and into formation. Might as well crank it up right now.
- All right, Master Chief. You guys heard him. Let's rock and roll.
- **Ready for** round two ?
- Hoo- yah, Master Chief.

Vocabulary Checkup

check out	확인하다
master Chief	선임 교관
awake	깨어있는
crank up	시작하다

Key Expressions

- Since the whole world seems to be awake at 0500, **why don't you** get your class out here and into formation. Might as well crank it up right now.
- **Ready for** round two?
- Hoo- yah, Master Chief.

Let's Review!

A: Look at those shoes. They look nice.

B: Do you think so? Hmm...

A: **Why don't you** try them on?

B: Yeah, I think I will. Okay, let's go in.

READING

Pre-Exercise Briefing

This briefing is to inform you of our logistics plans for the exercise in June next year. As you all know, the Brigade is going to go to Germany to practise armoured and infantry tactics. We, in the logistics group, will support them during the whole exercise.

First of all, fuel. The 16th supply unit will deliver all fuel necessary to the infantry and must do this once every three days. They will also be responsible for making sure that all the infantry units receive enough food during the exercise. The Field ambulance company are going to be responsible for medical cover for the whole exercise, although serious injuries must go by helicopter to the Field hospital because it will take too long for the field ambulance vehicles to drive there.

The 22nd Logistic Regiment will be responsible for all of the supplies to the armoured units, including the two tank squadrons that are stationed near the exercise area all year round. If vehicles break down, each regiment is responsible for taking their own vehicles to area 333 where a recovery vehicle will take them back to the main workshop for repair. 22nd Logistic Regt will also supply any other requirements to the infantry for the entire exercise.

The exercise starts on the 12 June and finishes on the 22 June. At the end of the exercise, units will return to their bases. The first to leave the exercise area will be the infantry, then the armour and finally the logistics troops. Please come to see me if you have any questions regarding this general outline.

Now I want to talk about the areas themselves. As some of you know, last year we had quite a few problems with the local residents in this area. I must point out to you all that this is an agricultural area and there are many farms. These farms grow all sorts of crops including potatoes, sweet corn and sugar beet. There also many others in the area that breed livestock – dairy cows, cows for market, pigs, chickens and various other animals for sporting purposes such as horses. These farms are totally out of bounds to all military vehicles, apart from specific logistics vehicles that have permission to go to these farms and collect meat and vegetables for the exercise. There are already contracts with farmers who will supply these products, so there is no need to approach anyone else.

The problems we had last year were very embarrassing for the army and the locals themselves. On one occasion, a troop of tanks went straight through the middle of three potato fields and then stopped to collect as many potatoes as they could. In another incident, two tanks drove straight through a farmer's barn and demolished it. W dround that the first incident had made the farmer very angry indeed and when our exercise damage control people visited him to arrange for compensation for the damage, he pulled a shotgun on one Captain and threatened to shoot him. Fortunately, he eventually laid down the weapon.

The second incident was much different. The Damage Control Team discovered the farmer was very happy indeed as he had asked the tank crews to destroy the barn for him as it was in terrible condition and he needed a new one He was even happier that an old plough horse died of shock during the incident and the farmer made a large profit in compensation. He had

enough money for a new barn, a new horse and a brand new tractor. The tank crew got a case of whisky each for the job. The crew, I should add, were all later arrested by military police.

Gentlemen, the Damage Control Team are under orders to arrest anyone causing damage to any property. If the damage was caused deliberately, the soldiers concerned will go to jail and then must leave the army. So, is everybody clear on this? {Pause} Good. Thank you.

Task

You may use any materials available in delivering your briefing - Poster paper, transparencies, rods and buttons, photographs etc.

Tips for making power point

- Don't bore by power-point.
- What you have to say is more important than the style of the presentation.
- Put your yourself in your listener's shoes.
- Words carry very little meaning, body language carries almost every thing.
- Keep it short and simple.

* Sample reproduced from Military English

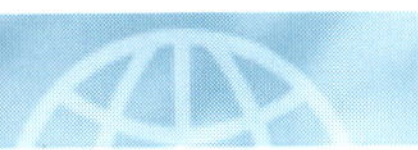

-Contents- 1

A. Non-Commissioned Officer
B. Company grade officer
C. Field grade and general officer
D. Positions
E. The Platoon Sergeant
F. Abbreviation

A. Non-Commissioned Officer 2

At the end of basic training, you will have lots of different tests and exams, and if your grades are good enough, you will go to **NCO** and technical training school to learn how to be a successful **NCO**.

3

원사	Sergeant Major(SGM)	
상사	Master Sergeant(MSG)	
중사	Sergeant First Class(SFC)	
하사	Staff Sergeant(SSG)	

B. Company grade officer 4

대 위	Captain(CPT)	
중 위	First Lieutenant(1LT)	
소 위	Second Lieutenant (2LT)	
준 위	Warrant Officer(WO)	

C. Field grade and general officer 5

대 령	Colonel(COL)	
중 령	Lieutenant Colonel(LTC)	
소 령	Major(MAJ)	

6

대 장	General(GEN)	
중 장	Lieutenant General (LG)	
소 장	Major General(MG)	
준 장	Brigade General(BG)	

D. Positions

7

주임원사	Command Sergeant Major
중 대 장	Company Commander
행정보급관	First Sergeant
소대장	Platoon Leader
부소대장	Platoon Sergeant
분대장	Squad Leader
반장	Section Leader
팀장	Team Leader

8

참모총장	Chief of Staff
사령관	Army Commanding General
군단장	Corps Commanding General
사단장	Division Commanding General
여단장	Brigade Commander
연대장	Regiment Commander
대대장	Battalion Commander

E. The Platoon Sergeant

9

The Platoon Sergeant must know all of the duties of the Platoon Commander. When the Platoon Commander is absent, he commands the platoon. He is responsible to the Platoon Commander for the day-to-day administration of the platoon.

F. Abbreviation

10

- 헌병: MP(Military Police)
- 전초: GP(Guard Post)
- 일반초소: GOP(General Outpost)
- 야전교범: FM(Field Manual)
- PT체조: Physical Training
- 군대용 원통형의 큰 자루: Duffle bag

11

- 군번 줄: Soldier's identification tag
- 군견: Military working dog
- 외박: Overnight pass
- 배 멀미: Seasickness
- 항해: Navigation
- 이임식: Departure Ceremony

12

- 야전상의: Field jacket
- 사격술예비훈련: PRI
(Preliminary rifle instruction)
- 국기게양: Flag raising
- 군악대: Military band
- 최전선: FEBA
(Forward Edge of Battle Area)

13

- ▶ 비 전투원 후송 작전: NEO
(Non-Combatant Evacuation Operation)
- ▶ 북방한계선
NLL (Northern Limit Line)
- ▶ 대량살상무기
WMD(Weapon of Mass Destruction)
- ▶ 전입 / 전출 PCS(permanent charge of station) in / PCS out

14

- 전사: Killed in Action
- 실종: Missing in Action
- 부상: Wounded in Action
- 전방이동: Onward Movement
- 포로: Prisoner of War
- 대외비: For Official Use Only
- II급 비밀: Secret
- I급 비밀: Top Secret

An English poem

The Waste Land

T. S. Eliot(1888-1965)

April is the cruelest month, breeding
Lilacs out of the dead land, mixing
Memory and desire, stirring
Dull roots with spring rain.
Winter kept us warm, covering
Earth in forgetful snow, feeding
A little life with dried tubers.

Appendix

DECK GROUP I
BOATSWAINS MATE
QUARTERMASTER
SONARMAN
RADARMAN
ORDNANCE GROUP II
FIRE CONTROL TECHNICIAN
MINEMAN

1. 작계표현

▶ 추가적인 전력이 연합사에 가용하게 될 것입니다.

Additional forces will be made available to CFC.

▶ 연특사는 공구사의 작전을 지원하기 위해 전략정찰을 실시할 것입니다.

CUWTF will conduct SR in support of ACC.

▶ 해구사는 한국해군과 미국해군으로 편성 될 것입니다.

NCC will be formed from ROK and US Navies.

▶ 연심사는 데프콘2 선포이후 창설 될 것입니다.

CPOTF will be activated upon declaration of DEFCON 2

▶ 후속부대는 연합사에 작전 통제 될 것입니다.

Follow-on forces will be OPCON to the CFC

▶ 이 결정으로 인해 연합사령관은 다양한 방책을 검토할 수 있도록 해줄 것이다.

This decision will allow CDR, CFC to review various COAs.

▶ 연합사령관은 지구사령관에게 권한을 위임하였다.

CDR, CFC delegated authority to CDR, GCC.

▶ 작계를 발전시키다.

The CFC staff participated in the development of this OPLAN (in developing this OPLAN)

▶ 본 작계는 전쟁의 재발을 억제하기 위해 작성되었다.

This OPLAN is designed to deter the resumption of hostilities.

▶ 본 작계는 상황에 따라 조정 될 수 있다.

This OPLAN is subject to modification in the context of a specific contingency.

▶ 본 부록은 사항을 기술 한다 ~ .

This Annex prescribes (describes, stipulates)~

▶ 병력보충의 우선 순위를 정하다.

Set priority over (prioritize) personnel replacement.

▶ 이 사항은 한국측 계통을 통해 요청 될 것입니다.

This request will be made through ROK channel.

▶ 절차를 준수하다.

Strictly adhere to procedures = Comply with procedures

▶ 전력을 통합하다.

Promote unity of efforts

▶ 신속히 추진하다.

Expedite (fasten, speed up) the process.

▶ 적에게 하도록 강요하다~ .

CFC must force the N,K attack to culminate rapidly.

▶ 능력 등을 보장하다~(   ) .

Ensure the ability to conduct expeditionary operations

▶ 저지 실패시

If deterrence fails~

주요약어

▶ 할 준비를 할 것 ~ BPT (be prepared to)

- 공통작전 상황도 COP (Common Operational Picture)
- 미정 TBD / TBA (to be determined (announced))
- ~에 의하면 IAW (in accordance with)
- ~하기 위하여 IOT (in order to)
- ~의 부근에 IVO (in vicinity of)
- 초기대응반 IRC (Initial Response Cell)
- 중간검토 IPR (in progress review)
- 담당부서 OPR (Office of primary responsibility)
- 협조부서 OCR (Office of collateral responsibility)
- 임무과제목록 METL (Mission Essential Tasks List)
- 공문 MOI (memorandum of instruction)
- 합의각서 MOA (memorandum of agreement)
- 준비명령 WARNO (Warning order)
- 전시교전규칙 WROE (wartime rule of engagement)
- 위기조치예규 CASOP (Crisis Action Standard Operating Procedures)
- 작전예규 OPSOP (Operations Standard Operating Procedures)

2. 주요회의 및 업무 간 표현

- ~ 하는데 제한 사항이 있다. : be limited in

  We are limited in conducting this operation.
- 로 국한되다. : be limited to (be confined to)

  Current WTC will be limited to in-garrison training
- 퇴역하다 VS 전역하다.

  retire VS get out
- 인쇄복사본/ 컴퓨터파일복사

  hard copy VS e-copy

- 본 지시사항등을 ( )에 반영하다 ~ .

  Incorporate CDR's guidance into this plan

  = implement ~ into~

  = integrate ~ into~

- 자료를 정리하다/통합하다.

  Compile (integrate, put together) comments from each staff

- 음주운전 DWI (driving while intoxicated)

- 보고는 영어로 진행되며 헤드셋을 통해 한국어로 동시통역이 제공되겠습니다.

  Briefing will be conducted in English with Korean simultaneous translation on the headsets

- 에 대해 감사의 뜻을 전합니다~ .

  I want to convey my sincere gratitude for~

- 모든 관계관에게 각 메일의,(서두에)

  ALCON (all concerned)

- 로부터 인가를 받다~ .

  Receive authorization from

- 저는 당신과 지휘관계에 있지 않습니다.

  I am not in your chain of command.

- 보고서를 상세히 검토하다.

  Review (go over) report in detail

- 추가적인 정보를 원하시면 작전예규를 참조 하십시오.

  Please refer to OPSOP for additional information.

- 현재로서는 이에 대하여 더 이상 가용한 정보가 없습니다.

  There is no further information available on this.

▶ 계속해서 업무를 추진하다.

We need to work on this

= go ahead with this work

= proceed with this work.

▶ 제가 작전처장 직무대리를 하고 있습니다.

I am representing the/ here on behalf of the Chief of operations

▶ 그 사람 너무 한 것 아니야?

He has gone over the limit

= He went too far

▶ 사명감

Sense of duty

▶ 과장으로부터 업무의 과실로 많은 질책을 받다.

I got slammed by chief of branch

▶ 현안업무보고

STU (Special Topic Update)

▶ 후임자/전임자/최근전입자

Replacement/predecessor/new comer

▶ 높은 수준의 전문성을 과시하다.

Demonstrate high standards and professionalism

▶ 로 구분되다~ .

Be divided into

▶ 누가 임석상관 이십니까?

Who's the senior?

= Who is hosting this meeting?

= Who will preside over?

▶ 제가 작참부에서 대표입니다.

I am representing C3

= I am a representative of C3

▶ 회의 시 지참물은 ~입니다 .

Deliverables are ...

▶ 누가 업무를 주관합니까?

Who will take the lead?

▶ 추후 확인 후 보고 드리겠습니다.

(회의 도중 잘 모르는 내용의 질문이 나왔을 경우)

I will get back to you.

▶ 그는 정말 훌륭한 장교다.

He is squared away

▶ 예산상의 제한

Budget constraints

▶ 전개현황은 슬라이드에서 보시는 바와 같습니다.

Deployment status is as shown on the slide.

▶ ~을 강조하고 싶습니다.

I would like to stress the importance of

= place an emphasis on~

▶ 단점들을 보완하다.

Make up shortfalls

▶ ~하는데 지속적으로 노력을 기울이다.

Continue efforts to~

▶ 2일간 쉬다.

I am going to take 2 days off.

▶ 실무자에게 확인 해봐야 하겠습니다.

I need to check with the actions officers.

▶ 요구를 충족시키다.(부응하다)

Meet requirements

▶ 취임을 축하하다.

Congratulations on assuming your position

▶ 나를 짜증나게 하다.

It pissed me off.

= I am sick of it

= I am fed up with it.

▶ 후속조치

Follow up measures.

▶ 정보공문을 전파하다.

Disseminate MOI to~

Distribute MOI to~

▶ 정보를 공유하다.

Share information between A and B

▶ ~ 와 협조하여

In collaboration with

= in coordination with

= in step with

= in line with

▶ 여러분들의 노고를 치하합니다.

I fully appreciate all of your hard work.

▶ 따뜻한 격려의 박수를 보냅시다.

Let's give him warm applause.

3. 공병분야

▶ D+1일차 상황보고 드리겠습니다.

I will brief D+1 situation report.

▶ 현재시간 까지의 주요조치 사항입니다.

The followings are major actions taken so far.

▶ FEBA B,C 지역의 장애물중 금일06:00시까지 군은 A군 지역은 78%, B 88% 설치 완료 되었으며 12월31일 12:00까지 100% 완료될 예정 입니다.

A corps has completed 78% and B corps area 88% of the obstacles in FEBA "A" and "B" areas as of 0600 today. Also all obstacles in those areas will be completed by 31 Dec 1200.

▶ 주요 관심사항으로 연합사 차후작전을 위해 C군에서 B군으로 10 MGB 개 셀이 12,30 12:00 일까지 전환이 완료되겠습니다.

The main focus is on the transfer of 10 MGB sets from C Corps to B Corps for CFC's future operation as of 30 Dec1 200.

▶ 이상입니다.

That is all for CFEN Sir.

▶ 댐 수문 및 통제시설 거부시는 필요한 핵심부품을 제거/파괴 해야 합니다.

When we deny dams and control facility, remove or destroy main and key part.

▶ 상급부대로부터 거부권한이 위임되면 지휘관의 승인하에 우선순위에 의거 거부표적을 거부합니다.

When denial authority is delegated from higher units, deny targets according to the priorities with the commanders approval.

▶ 부교 구축시 고려사항은 유속 과 하폭 그리고 수심입니다.

We have to consider velocity of current, width and depth of river when we plan to set up floating bridge.

▶ 기동 부대 부지휘관이 도하지역 지휘관으로 임명됩니다.

Deputy commander of maneuver unit is appointed as a commander of river crossing area.

▶ 지뢰지대 계획은 화력, 관측 및 다른 장애물 계획과 연계 되어야 합니다.

A mine field plan should be linked with fire, observation and other obstacle plans.

▶ 금일 회의에 참석하신 미 육군 공병감, 공병 사령관 그리고 000 주한미군 및 8군 공병 관계관 여러분께 감사드립니다.

I'd like to give my warm welcome to the US Army Chief of Engineer, the Commander of 000 ENCOM and all engineer officers from USFK, EUSA.

▶ 금일 토의는 GATOR 지뢰 설치 요청 절차개선을 위한 공병 실무자 모임입니다.

Today's meeting is for everyone involved to improve the GATOR minning request procedure.

주요약어

▶ 한국 장애물 체계 KBS (Korean Barrier System)

▶ 거부표적 Denial target

▶ 구축장애물 Constructed Obstacles

▶ 급속장애물 H asty Obstacle area

▶ 낙석 Rock Falls /Slide drops

▶ 대 전차구 Anti-Tank Ditch

▶ 저성능/고성능폭약 Low/High Explosive

▶ 권한위임 Authority Delegation

▶ 자폭기능 Self Destruction Capability

- 상자형 지뢰 MOPMS (Modular Pack Mine System)
- 광역지뢰 WAM(Wide Area Munition)
- 지뢰 제거 로라 TWMP (Track Width Mine Roller)
- 간편 조립교 M GB (Medium Girder Bridge)
- 교량전차 AVLB (Armored Vehicle Launched Bridge)
- 부교 FB (Floating Bridge)
- 강습부교 AFB (Assault Floating Bridge)
- 다목적교량중대 MRBC (Multi-Role Bridge Company)
- 만수위 FWL (Full Water Level)
- 저수위 LWL (Low Water Level)
- 제한수위 RWL (Restricted Water Level)
- 군 건설공사계획 MILCON (Military Construction Program)
- 지상유류분배체계 IPDS (Inland Pipeline Distribution System)
- 해안유류분배체계 OPDS (Offshore Pipeline Discharging System)
- 경의선 Western Transportation Corridor
- 동해선 Eastern Transportation Corridor
- 연합 지형 분석실 CTAT (Combined Terrain Analysis Team)
- 비행장 피해복구 ADR (Air Field Damage Repair)
- 항공단말보급소 ATSP (Air Terminal Supply Point)
- 주한미군 공병참모부 FKEN (Forces Korea Engineer)
- 공병감 COE (Chief of Engineer)
- 공병통제소 ERP (Engineer Regulating Point)
- 전투공병야외훈련 FTC (Field Training Course)
- 원격조종장치 RCU (Remote Control unit)
- 교통통제소 TRP (Traffic Regulation Point)
- 시설대 DPW (Directorate Public Work)

4. 정보 분야

▶ 북한의 전투력에 대해서 보고하겠습니다.

I will brief you on NK's combat strength.

▶ 먼저 약0명의 북한군이 00에 배치되어 있으며 또, 약 0명의 현역과 예비군을 보유하고 있습니다. 그리고, 북한이 보유하고 있는 무기들은 0대의 탱크, 0대의 장갑차,0대의 다연장 로켓포, 0대의 포병화기, 0대의 방공포 병화기 등이 있습니다.

To begin with, there are 0 NK soldiers located 00, and its ground troops are estimated at 0 personnel on active duty and reserves. Other NK's Army weapon assets are 0 tanks, 0 APCs, 0 MRLs, 0 artillery pieces, and 0 ADA guns.

▶ 그 중에는 약0명 규모의 특수작전부대가 있습니다. 이들의 임무는 정찰, 정규전지원, 적 후방에 제2전선 형성, 북한내 적 특수작전부대에 대응 및 국내 안보 유지 등 입니다.

Of these troops, about 0 are Special Operation Forces, whose missions are to conduct reconnaissance, to support conventional operations, to establish a second front in the enemy's rear area, to counter enemy SOF in NK, and to maintain internal security.

▶ 다음은 북한의 해군과 공군의 전투력입니다. 먼저 해군기지의 위치와 전투력입니다. 북한의 함대사령부는 0에 위치하고 있고, 해군기지들은 0에 위치되어 있습니다. 그리고 북한 해군 병력은 약0명이며, 0척 이상의 함정을 보유하고 있습니다. 그리고 이들은 해상 교통로에 위협을 줄 것입니다.

Next is the combat strength of NK's Naval and Air Forces. First the location and strength of NK's Naval Organization. Fleet

Command Headquarters are located at 0, and Naval Bases are located at 0. There are 0 personnel and 0 plus vessels. These are all a threat to sea lines of communication.

▶ 계속해서 북한 공군에 대해 보고 드리겠습니다. 북한의 공군 병력은0명이며, 0대의 비행기를 보유하고 있습니다. 북한은 육·해·공군 이외에도 전구 탄도미사일, 화학무기 그리고 핵 능력 등을 이용 전쟁 발발시 대규모 사상자를 유발시킬 수 있습니다.

Following is infomation regarding NK's Air Force. There are about 0 personnel and 0 aircraft. Other than NK's Army, Navy and Air Force, there are theater ballastic missiles, chemical weapons, and nuclear capabilities which could result in a great number of casualties.

주요 약어

- 집단군 포병군 AAG (Army Group Artillery Group)
- 군포군 CAG (Corps Artillery Group)
- 사포군 DAG (Division Artillery Group)
- 정보참모부부장 DDI (Deputy Director of Intelligence)
- 조선민주주의 인민공화국 DPRK(Democratic People's Republic Of Korea)
- 전자 정보 ELINT (Electronic Intelligence)
- 도하공병 ERC (Engineer River Crossing)
- 전선사 기동예비 FCMR (Front Command Mobile Reserve)
- 총참모부 GSD (General Staff Dept)
- 징후경보 I & W (Indicator & Warning)
- 영상정보 IMINT (Imagery Intelligence)
- 정보 보안사령부 INSCOM (Intelligence & Security Command)

- 미태평양 합동 정보센타 JICPAC (Joint Intel Center Pacific)
- 인민군 KPA (Korean People's Army)
- 노동당 KWP (Korean Worker's Party)
- 경보병 여단 LIB (Light Infantry BDE)
- 기계화보병 여단 MIB (Mechanized Infantry BDE)
- 방사포 MRL (Multiple Rocket Launcher)
- 주요사태목록 MSEL (Master Scenario Event List)
- 특이사항 없음 NSTR (No Significant To Report)
- 지대공 유도탄 SAM (Surface To Air Missile)
- 이동발사대 TEL (Transporter Erector Launcher)
- 지하갱도 시설 UGF (Underground Facility)
- 접근로 Avenue of Approach
- 기동회랑 Mobility Corridor
- 횡단지점 Crossing Site
- 집결지 Staging Area
- 공격 대기지역 Assembly Area
- 공작모선 Mother-ship (Command Ship)
- 예방책 Preliminary measures = protective measures
- 중간 정도의 활동 At moderate levels
- 표적추적 레이다 Target tracking radar
- 꽃게잡이 철에 In crab season

5. 작전분야

▶ 지난 보고이후 1개의 주요 상황이 발생하여 CCIR 14번이 황색상태로 격상되었습니다.

There was one significant event which activated (triggered) CCIR # 14 yellow since the last update.

▶ 보고드릴 특이사항이 없습니다.

There is nothing significant to report (NSTR)

▶ 질문이 없으시면, 이어서 정참부 보고 드리겠습니다.

Pending (subject to) your questions, I will be followed by C2.

▶ 모든 야전군은 경계태세C형을 유지하고 있으며 일부 표시된 부대만 B형을 유지하고 있습니다.

All field Armies are at readiness posture C with selected units at B.

▶ 한국군 전 함대는 해상경계 태세 및 대잠전 대비태세 C형을 유지하고 있습니다.

All ROK fleets are posture for anti-coastal and maintaining Anti submarine warfare (ASW) readiness "C".

▶ 모든 비행장, 패트리어트 포대는 작전 가능합니다. (정상임무 수행중입니다.)

All airfields and patriot missiles/batteries are operational (fully mission capable).

▶ 지난 보고이후 발령된 전문 및 명령은 없습니다.

There is no messages or orders released since the last update.

▶ 주요 일정에는 1개의 변동사항이 있으며 청색으로 표시되어 있습니다.

There is one change to the significant events timeline which is highlighted in blue on the screen.

▶ 본 훈련은 전입자로 하여금 위기조치 예규 절차를 숙지 할 수 있도록 계획되었습니다.
This training is designed to familiarize new personnel with the CASOP procedure.

▶ 주한미군/연합사 인근에서 예정된 시위는 총 10개입니다.
There are 10 civil demonstrations/disturbances scheduled near the USFK/CFC installations.

▶ 경의선 통과 현황은 공통작전 상황도 상에 표시되어 있습니다.
MDL crossings at Western Transportation Corridor are depicted

▶ 1번 카메라를 확대하고 공통 작전 상황도에 시야를 표시 하여 주십시오.
Please blow-up/expand the camera # 1 and show us the field of view on the COP.

▶ 그들은 주한미군 철수 및 소파(SOFA) 개정을 요구할 것으로 예상됩니다.
They are expected to demand USFK to leave Korea and SOFA revision.

▶ 인원의 통보 및 소집
Notification and Recall of personnel

▶ 위기 조치반을 운용하다.
CAT will be stood up (activated)
= CFC will stand up (activate) CAT

▶ 초기대응반을 소집하다.
CJ3 Ops will instruct EOC to muster (recall) IRC in a timely manner.

▶ 정전상황에서 전시상황으로 전환되는 동안
During the transition from armistice to wartime

▶ A지휘소에서 B지휘소로의 전환이 용이하도록 절차를 발전시키다.
Develop procedures to ensure smooth transition from CP A to CP B.

▶ 주요상황일지를 유지하다.

Watch officer will maintain significant events log.

▶ 상황이 전개됨에 따라

As the situation dictates

= as situation develops to crisis

= as situation continues to develop

▶ 상황이～할 시간을 허용하지 않다.

Circumstances do not allow time to ~.

▶ 긴박한 상황에서

In a time-sensitive situation

= In case of emergency

▶ 시간이 가용치 않을 경우

If time is not available

= If time does not permit

= If there's no time

▶ 전문하달 절차

Message Release Procedures

▶ 이 조치는 한국 항공기를 비상대기 발령시킬 것입니다.

This measure will place ROK aircraft on alert.

▶ 모든 인원은 본 예규상에 명시된 용어를 숙지하여야 합니다.

All personnel must be familiar with the terms listed in this regulation.

▶ 전방 지역으로 투입되다.

Units are committed to forward area.

▶ 전방 지역으로 전개되다.

Units are deployed to forward area.

▶ 대 화력전 수행본부의 창설 및 해체

Establishment and dis-establishment of counterfire HQs

▶ 전투력이 현저히 저하되다.

PDC has been attrited.

We've got damaged severely.

▶ 공중 탐지 능력을 상실하다.

We've lost (air detection) capability.

▶ 전투력은 15% 수준이다.

Combat power (combat strength) is at 15 %.

▶ 전투력의 15%가 감소했다.

Combat power has decreased by 15 %.

▶ 전투력이 15% 수준으로 감소했다.

Combat strength is reduced (decreased) to 15 %.

▶ (전투력을) 할당하다.

CFC will assign aviation assets to GCC.

▶ 전투력을 복원 중이다.

This company is reconstitutioning.

= Unit is on-going reconstitution.

= Currently, TROKA has 5 Divisions undergoing reconstitution.

▶ 요망된 효과를 얻지 못했다.

Desired effects have not been realized
(achieved, obtained).

▶ 계획이 순조롭게 진행중이다.

We are on plan.

▶ 계획이 예상보다 빨리 진행중이다.

We are pretty ahead of plan.

▶ 적의 계획 실행을 방해하다.

Keep (throw) the enemy off his war plan.

▶ 한라선 이남에 강력한 방어 진지를 형성하다.

GCC Established defensive strong points/line along South of Halla line.

▶ 반격작전을 준비하다.

CFC is preparing for counter- offensive OPS.

▶ 돌파를 거부하다.

Deny penetration.

▶ 돌파구 확장을 저지하다.

Penetration is contained with forces.

▶ 1군은2군단지역의 FEBA A를 돌파된 상태에서 계속 방어중이다.

FROKA is defending along FEBA A with a penetration of FEBAA in 2nd Corps area.

▶ 적의 공격기세가 약화되다.

NK's attack is slowing down.

= NK's attack got bogged down.

▶ 전방 초월작전을 실시하다.

CMEF will conduct FPOL of (through) TROKA.

▶ 병행공격하다.

FROKA will continue to attack NK in parallel with 00 Corps.

= attack north with 00 Corps abreast

= launch a parallel attack

주요약어

▶ 복합체계 분석

SOSA (System of system analysis)

- 효과중심작전

  Effects Based Operations

- 사령관 결심 시행 주기

  PDE cycle (Planning, Decision and Execution)

- 동맹군 전쟁 예비 물자

  WRSA (War reserve stock for allies)

- 서북도서

  NWI (Northwest Islands)

- 대 화력전 임무 수행 부대

  CFTF (Counterfire Task Force)

- 전방초월

  FPOL (forward passage of line)

- 작전가능

  FMC (fully mission capable)

- 임무수행 불가

  NMC (non mission capable)

- 상황조성전문

  SDM (scenario development message)

- 경의선/동해선

  TC (transportation corridor)

- 현안 업무 보고

  STU (special topic update)

- 상황보고서

  SITREP (Situation Report)

- 미 본토

  CONUS (continental US)

▶ 특이사항 없음

NSTR (nothing significant to report)

▶ 대잠전

ASW (anti-submarine warfare)

▶ 비전투원 후송 작전

NEO (Non-combatant Evacuation Operations)

= "Courageous Channel"

▶ 한강하구

HRE (Han River Estuary)

▶ 주요사태목록

MSEL (Master scenario events list)

▶ 최적 침투시기

OIP (optimal infiltration period)

▶ 사령관 판단서

CDR's Estimate

▶ 전술유도 무기체계

TACMS (Tactical Missile System)

▶ 전개준비명령

PTDO (Prepare to deploy order)

6. 군수 분야

▶ 군수 기반시설은 황색상태입니다.

Logistics infrastructure is amber.

▶ 미 전구 군수기지는 ○○,○○,○○에, 한국육군 군수사령부는 ○○에 위치하고 있습니다.

The U.S theater logistics bases are established at ○○, ○○, ○○. The ROKALC is headquartered in○○.

▶ 한·미간 군수지원을 협조하기 위한 연합군수 협조기구들은 정상적으로 운영 중입니다.

All combined logistics coordinating agencies are operational and prepared to coordinate combined logistics issues.

▶ WRSA-K 물자는 한국군으로 이양되어 불출 중이며, 24~48 시간 이내 완료될 예정입니다.

The issue of WRSA-K stocks to ROK is underway. Anticipate completing the issue in the next 24~48hours.

▶ 전반적인 군수지원은 긴요 탄약을 제외하고는 작전지원이 가능합니다.

Assess that logistics support will continue to meet operational requirement with some limitations, primarily in preferred ammunitions.

▶ 적 특작부대 및 스커드 공격이 주요 군수기반시설에 계속 되고 있습니다.

NK SOF and SCUD activity continue to target key logistics facilities.

▶ 몇 가지 제한사항이 있지만 반격작전을 위한준비를 지원 할 수 있습니다.

We can support preparation for counter attack with some limitation.

▶ 공군사전배치선박 메릴린호가 C-○일에 진해항에 입항 하였습니다.

MV Merlin, air force munitions preposition ship, has docked at Jinhae port at C-○.

▶ 미8군사 MLRS 탄약재고는 소요량 대비 초과보유하고 있으며, 이양 요청시 지원이 가능할 것입니다.

Eighth United States Army stockpiles of MLRS exceed requirements and may be able to support a transfer request.

▶ 대 화력전 에는 어떤 탄약이 사용됩니까?

What kind of munitions do you use in Counterfire operation?

▶ 보급품종별 지원능력은 녹색상태입니다.

All classes of supply are green.

▶ 전반적인 군수 상황은 어떻습니까?

What is the overall logistics posture?

▶ 연합군수운영본부는 CC서울 또는 탱고지휘소에 위치 할 것입니다.

The CLOC will be located at CC Seoul or CP Tango

▶ 연합군수지원은 자국군 지원이 원칙입니다.

Combined logistics support is based on the principle that each nation supports its own forces.

▶ 연합사의 군수지원 지속 능력 목표는 00일 이상입니다.

CFC objective for Logistics sustainability is above 00 days.

▶ 연합사 차원의 군수 지원은 최종방책으로 운용될 것이다.

CFC level logistics support will be used as a last resort.

▶ 이두항만의 정리능력은 문제가 될 것 같습니다.

Port clearance will likely be a problem for these two ports.

▶ 전반적인 탄약 상태는 녹색이나, D+○일에 황색으로 변경될 예정입니다.

Overall ammo status is Green. But it will be Amber at D+○.

▶ 전쟁예비 사전비축재고는 고갈 되었습니다.

Prepositioned War Reserve Stocks are exhausted.

▶ 전체적인 공군탄약은 금일현재 ○일분을 보유하고 있습니다.

As of this date, there are○days of the overall Air forces munitions remaining.

▶ 현재진행 중인 지상군정비 지원능력은 어느 정도(%)입니까?

What is the current level of GCC maintenance support capability?

▶ 현재 해군 탄약은 며칠분입니까?

How many days do munitions last for the entire naval forces?

= How many days can you sustain on your munitions level?

▶ 군 통제 선박 운용단은 언제 창설됩니까?

When will the Military Control Vessel Operation Group established ?

▶ 긴요 탄약

Preferred munitions

▶ 대량 유류소요

Bulk petroleum requirement

▶ 미군이 사용 중인 주요장비의 제원에 대하여 알고 싶습니다.

I want to know the specifications of main USFK equipment.

▶ 미군 전용탄약의 한국군 이양

Transfer of US-intended munitions to ROK forces

▶ 부대별 기본휴대량

Unit basic load stocks

▶ 연합사 군수 참모부는 ○개처 ○○개과로 구성되어 있습니다.

C4 CFC is composed of ○○branches organized under ○ divisions.

▶ 유류능력보고서는 누가 담당 합니까 ?

Who is the POC for POLCAP ?

▶ 육로와 철로에 관한 일반제원을 주실 수 있습니까?

Can you provide information regarding the highways and railroads data?

▶ 주한미군의 정비지원시스템은 어떻게 됩니까?

How does USFK's maintenance support system work ?

▶ 목포항의 야적장과 저장소 능력에 관한자료를 구할 수 있습니까?

Can I get information concerning the open storage and covered storage capability of the port of Mok-po ?

▶ 항만 처리능력

Seaport throughput capability

주요 약어

▶ 공중기동사령부

AMC (Air Mobility Command)

▶ 대외군사판매제도

FMS (Foreign Military Sales)

▶ 병참선

LOC (Line Of Communication)

▶ 소요보급률

RSR (Required Supply Rate)

▶ 연합공수사무소

CALO (Combined Airlift Office)

▶ 연합 지상군 군수위원회

CGLC(Combined Ground Logistics Committee)

▶ 연합 군수 협조기구

CLCA(Combined Logistics Coordinating Agency)

▶ 연합 항만 협조본부

CSCC(Combined Seaport Coordination Committee)

▶ 연합 수송 위원회

CTB (Combined Transportation Board)

▶ 이동 중 식별체계

ITV (In-Transit visibility)

▶ 정밀 유도폭탄

PGM (Precision Guided Munitions)

▶ 전시 이동계획서

WMP (Wartime Movement Program)

▶ 전시 지원

WHNS (Wartime Host Nation Support)

▶ 지상유류분배장비

IPDS (Inland Petroleum Distribution System)

▶ 지상주기대수

MOG (Maximum On Ground)

▶ 탄약보급소

ASP (Ammunition Supply Point)

▶ 통제보급률

CSR (Controlled Supply Rate)

▶ 한국 국적선박

KFS (Korean Flag Ship)

▶ 한국 육군 군수사령부

ALC (Army Logistics Command)

▶ 합동정밀직격폭탄

JDAM (Joint Direct Attack Munition)

▶ 항공추진보급기지

ATSP (Air Terminal Supply Point)

▶ 해상유류분배장비

OPDS (Offshore Petroleum Discharge System)

7. 기획 분야

▶ 안녕하십니까? 저는 기획처에 근무하는 김 중령 입니다.

Good moming ladies and gentlemen, I'm LTC Kim from C-5 Plans Divisions.

▶ 저희 지휘소를 방문하여 주신 여러분들을 전 부대 장병들을 대신하여 감사드립니다.

On behalf everyone assigned to our base/post, welcome to our Headquarters.

▶ 앞으로 30분간에 걸쳐서 기획참모부 편성과 임무에 관하여 설명 드리겠습니다.

We are pleased to present a briefing about our organizations and our role for next 30 minutes.

▶ 기참부는( )개 처와 행정과로 편성되어있으며 각처(과)는 한.미 Counterpart로 균등하게 편성함을 원칙으로 편성하고 있습니다.

C5 is composed ( ) major divisions and admin sections based on combined ROK-US organizational Structure.

▶ 행정과는 한·미 각각 편성되어 부장과 차장에 대한 보좌관 임무를 병행하여 수행합니다.

The Admin is responsible for staff and fills in as XOs to each country Assistant Chief of Staffs.

▶ 경청해 주셔서 대단히 감사합니다.

Thank you very much for your attention.

▶ 이상으로 브리핑을 마치겠습니다. 질문 있습니까?

Sir, subject to your questions, this concludes my briefing.

= Sir, this concludes my briefing. Do you have any question?

▶ 이 박사님, 감사합니다. 훌륭한 발표(연설/강의) 였습니다.

Thank you very much, Dr. Lee, for your excellent presentation (speech/lecture).

▶ 좋은 연설(견해를 피력/분석/평가)을 해주신 이 대령님께 감사드립니다.

Thank you COL. Lee for your excellent comment (view/analysis/assessment).

▶ 마이크를 사용해서 자기소개를 해 주십시오.

Please use the microphone and identify yourself.

▶ 질문한 가지만 더 받도록 하겠습니다.

Let's go only one more brief question, please.

= I will take one last question.

▶ 아직 시간이 많이 있으니 여러분의 견해나 제안을 마음껏 발표해 주십시오.

We still have more time for some questions, comments, and discussions.

▶ 질문이나 발표 하실 분 있습니까?

Are there any questions or comments?

▶ 다른 견해나 반박하고자 하시는 분이 계시면 말씀해 주십시오.

(질문이 없을 때)김 소령님, 더 이상 질문이 없는 것 같습니다.

수고하셨습니다.

Any other idea or different opinion?

(in no question) It looks as if you've covered the briefing completely, Major Kim, Thank you very much.

▶ 주제를 바꾸는게 좋겠습니다.

I would like to change the subject.

▶ 말씀을 막아서 죄송합니다만, 주제를 넘어선 언급인 것 같습니다, 김 소령님.

Major Kim, pardon me for interrupting you, we have to stay on the topic of discussion.

▶ 논점에서 벗어난 것 같습니다. 다시주제로 돌아오겠습니다.

We seemed to have gotten off the main point. We will have to go back to our topic.

▶ 질문이 있습니다.

I would like to ask a question.

= I have a question.

▶ 제 생각으로는 대부분의 참석자가 A와B의 차이점을 이해하지 못하는 것 같습니다.
자세하게 설명해 주시겠습니까?
I think most of us may not understand the distinction between A and B.
Could you explain them clearly?

▶ 이 박사님, A에 관한 박사님의 의견을 말씀해 주십시오.
Dr. Lee, I'd like to have your opinion on A.

▶ 매우 흥미로운 발표를 해주신 미스터 김에게 감사드립니다.
I appreciate Mr. Kim on a very interesting presentation.

▶ 매우 흥미로운 브리핑이었습니다. 그런데 두 가지 질문하고 싶은 점이 있습니다.
I really enjoyed your interesting briefing, but I have two more questions.

▶ 제 말씀이 답변이 되었습니까?
Did I answer your question?

▶ 질문의 답을 얻으셨습니까?
Did your question get answered?

▶ 5분 동안 휴식을 갖겠습니다.
Let's take a five minute break.

▶ 좀 더 크게 말씀해 주십시오.
Would you speak up please?

▶ 이제 끝낼 시간이 되었습니다.
We have reached the end of our time.

▶ 시간이 매우 빠르게 지나갔습니다. 벌써5시가 되었군요. 이제 회의를 마쳐야겠습니다.

Time has gone by so fast. It's already five o'clock. We have to conclude the meeting.

▶ 여러분, 이상으로 회의를 마치겠습니다.

Ladies and Gentlemen, I declare the meeting closed/adjourned.

주요 약어

▶ 조기경보통제체계

AWACS(Airbone Warning and Control System)

▶ 연합유관기관 협조단

CIACG (Combined Interagency Coordination Group)

▶ 전쟁지휘

COCOM (Combatant Command)

▶ 연합권한위임사항

CODA (Combined Delegated Authority)

▶ 공통작전상황도

COP (Common Operational Picture)

▶ 신속억제방안

FDO (Flexible Deterrent Options)

▶ 전투력증강

FMP (Force Module Package)

▶ 수도권

GSMA (Greater Seoul Metropolitan Area)

▶ 주둔국 지원협정

HNSA (Host Nation Support Agreement)

▶ 통합임무명령서

ITO (Integrated Task Order)

▶ 한국방공식별구역

KADIZ (Korean Air Defense Identification Zone)

▶ 한국작전구역

KAO (Korean Area of Operations)

▶ 병참선

LOC (Line of Communications)

▶ 전쟁이외의 작전

MOOTW (Military Operations Other Than War)

▶ 비전투원후송작전

NEO(Non-Combatant Evacuation Operation)

▶ 북방한계선

NLL (Northern Limit Line)

▶ 작전지휘

OPCOM (Operational Command)

▶ 작전통제

OPCON(Operational Control)

▶ 작전계획

OPLAN(Operational Plan)

▶ 작전명령

OPORD(Operational Order)

▶ 대량살상무기

WMD(Weapon of Mass Destruction)

▶ 수용가능성 Acceptability

▶ 적절성 Adequacy

▶ 가정 Assumption

▶ 비대칭 수단 Asymmetric means

- 전역 Campaign
- 중심 Center of Gravity
- III급비밀 Confidential
- 사후관리 Consequence Management
- 억제 Deterrence
- 타격작전 Direct Action
- 지시 Directive
- 최종상태 End State
- 시행가능성 Feasibility
- 대외비 For Official Use Only
- 주도권 Initiative
- 통합 Integration
- 상호운용성 Interoperability
- 전사 Killed in Action
- 실종 Missing in Action
- 무력화 Neutralization
- 전방이동 Onward Movement
- 포로 Prisoner of War
- 수용 Reception
- II급비밀 Secret
- 대기 Staging
- I급비밀 Top Secret
- 평문 Unclassified
- 부상 Wounded in Action

8. 통신 분야

▶ 우리는 유엔사/연합사 임무를 수행한다.

We conduct UNC/CFC's missions.

▶ 한국군 스파이더와 미군 전술 장비간 상호연동이 가능하다

ROK SPIDER switch is capable of interfacing with US tactical switche.

▶ 공통작전상황도를 작성하다.

Develop(build) a COP.

▶ 부대 위치를 지도에 표시하다.

Depict unit location on the map.

▶ 부가적인 통신지원을 제공하다.

Provide additional comm support.

▶ 통신계획은 생존성 제고를 위해 보호대책을 강구토록 계획한다.

Comm plans provide for physical protection to enhance survivability.

▶ C4시스템의 신뢰성과 생존성을 향상하다.

Improve C4 system reliability and survivability.

▶ 정보체계를 보호하고 방어하다.

Protect and defend information systems.

▶ 유엔사/연합사 정보체계를 관리하다.

Manage UNC/CFC information system.

▶ 주요 임무수행에 중점을 두다.

Focus on critical tasks.

▶ 주요 통신자산의 사용을 재할당하다.

Reallocates the use of critical commassets.

▶ 작전적 영향을 평가하고 결정하다.

Make a determination of operational impact.

▶ C4 상호운용성을 발전시키고 시행하다.

Develop and implement C4 interoperability.

▶ 의명, 지휘통제통신망을 구성하라.

On order, establish command control comm network.

▶ 전화번호, 전자우편 및 웹사이트 목록을 발전시키고 전파하다.

Develop and distribute telephone, email and website directories.

▶ 임무 및 책임을 수행하다.

Perform your mission and taskings.

▶ 같이 협조하여 통신부록을 만들어야 한다.

We need to work together to formulate comm annex.

▶ 이상황에 통전부 위기조치요원 이 적용한 위기조치항목은 무엇입니까?

What items of the crisis action measures were applied in this cases by the C6 CAT?

▶ 이회선은 요구시만 사용 가능합니다.

It is an on call circuit.

▶ 하드웨어와 랜 문제를해결하기 위해기술자를 파견하다.

Dispatch a technician to take care of H/W or LAN problem.

▶ 비용분담에 대한 실무회의를 개최하고자 합니다.

We want to hold a workshop on the cost sharing.

▶ 부대 소개준비를 해주시겠습니까?

Would you prepare a mission brief?

▶ 그것은 서버 문제가 아니라 회선 문제입니다.

It is not a sever problem, but a circuit problem.

▶ 통전부는 보다 융통성있는 통신망 제공을 위해 연구한다.

CJ6 studies to provide greater technical flexibility.

▶ 사업 진행이 부진한 이유는 무엇입니까?

What is the reason for this project's poor progress?

▶ 전화기 설치 이전이 필요하다.

Telephones need to be relocated.

▶ 파일이 게시되다.

A file is posted.

▶ 기술자가 C4I체계의 연결 여부를 시험한다.

Technician tests the connectivity of C4I system.

▶ 그가 통신망을 소통시켰다.

He activated circuits.

▶ 이 장비의 일반제원 및 운영개념.

The general characteristics and operational concept of this equipment.

▶ 우리는 고장개소를 정확히 파악해야 한다.

We must identify the reason for outage.

▶ 전문가가 VTC 통신망을 복구한다.

An expert is restoring the VTC network.

▶ 모든 사람은 운용 요령을 숙지해야한다.

All personnel should be familiar with operating procedures.

▶ 접속 불량 때문에

Due to poor connection

▶ 그가 추후 오해의 소지를 제거했다.

He avoided future misunderstanding.

주요약어

▶ 협조체계

JADOCS(Joint Automated Deep Operations Coordination System)

▶ 육군 전술 지휘통제체계

ATCCS(Army Tactical Command and Control System)

▶ 위기 조치반

CAT(Crisis Action Team)

▶ 통신전자 운용지시

CEOI(Communications-Electronics Operating Instructions)

▶ 전술용 비화전화

DSVT(Digital Secure Voice Terminal)

▶ 범세계 방송 서비스

GBS(Global Broadcast Service)

▶ 군주파수 관리위원회

JMFC(Joint Military Frequency Committee)

▶ 해군 전술데이터 체계

KNTDS(Korean Naval Tactical Data System)

▶ 군사위성

MILSTAR(Military Communication Satellite)

* **조규택**(문학박사)은 해군영어교관을 역임했으며
지금은 계명문화대학 부사관과에서 군사영어를 가르치고 있다.

군사영어
MILITARY ENGLISH

초판인쇄 2010년 3월 1일
초판발행 2010년 3월 5일

엮은이 / 조 규 택
펴낸이 / 이 승 엽
펴낸곳 / 도서출판 **정림사**
700-809
대구광역시 중구 대봉동 47-13
등 록 번 호 • 제03-01-340
대 표 전 화 • (053) 424-1661
팩 스 • (053) 424-1665

ISBN 978-89-6064-032-0 93740

〈값 14,000원〉